建筑立场系列丛书 No.46

C3

重塑建筑的地域性

Re-assessing Local Identity

汉英对照版

（韩语版第362期）

韩国C3出版公社 | 编

刘懋琼 陈玲 | 译

大连理工大学出版社

建筑立场系列丛书 No. 46

4 重塑建筑的地域性

004 在全球化建筑的背景下重塑建筑的地域性 _ *Angelos Psilopoulos*

010 瓦山旅社 _ Amateur Architecture Studio

028 Brockholes访客中心 _ Adam Khan Architects

040 Beautour博物馆和生物多样性研究中心 _ Agence Guinée Et Potin Architectes

052 福古岛自然公园总部 _ Oto Arquitectos

070 Sancaklar清真寺 _ Emre Arolat Architects

088 玻璃农场 _ MVRDV

100 do Morro竞技场 _ Herzog & de Meuron

118 水上的建筑 _ Álvaro Siza + Carlos Castanheira

140 持续且不断发展的传统

140 持续且不断发展的传统 _ *Paula Melâneo*

144 微胡同 _ standardarchitecture

156 雅法厂房 _ Pitsou Kedem Architects

166 石材浇筑的住宅 _ SPASM Design Architects

178 古梅斯苏别墅 _ Cirakoglu Architects

186 建筑师索引

C3 No. 46 Re-assessing Local Identity

Re-assessing Local Identity

004 *Re-assessing Local Identity in the Context of Global Architecture _ Angelos Psilopoulos*

010 Washan Guesthouse _ Amateur Architecture Studio

028 Brockholes Visitor Center _ Adam Khan Architects

040 Beautour Museum and Biodiversity Research Center _ Agence Guinée Et Potin Architectes

052 Fogo Island Natural Park Headquarters _ Oto Arquitectos

070 Sancaklar Mosque _ Emre Arolat Architects

088 Glass Farm _ MVRDV

100 Arena do Morro _ Herzog & de Meuron

118 The Building on the Water _ Álvaro Siza + Carlos Castanheira

Persistent and Evolving Traditions

140 *Persistent and Evolving Traditions _ Paula Melâneo*

144 Micro-Hutong _ standardarchitecture

156 Factory Jaffa House _ Pitsou Kedem Architects

166 The House Cast in Liquid Stone _ SPASM Design Architects

178 Gumus Su Villas _ Cirakoglu Architects

186 Index

重塑建筑的地域性

Re-assessing

建筑物和建筑之间存在着较大的差距，这种差距是一种距离，是后者为了证明自己不琐碎的特性而要跨越的距离。多数人将其误以为是矫饰，从而将美感作为令人赞叹的建筑的最重要的指标。建筑师通常宣扬一种诠释性立场，反对任何一个给定的情景，因为在建筑的形式或功能中，给定的场景通常是在先的，或者它所占据的环境是前提，这看起来好像要达到的主要建筑要求是提高其实用性，远离常规性。

我们这里所回顾的例子都是与背景相对立的。建筑师除了对环境不加以重视外，对所有的情况都加以规划，他们提出与环境条件相关的定义元素的术语。这不是你的学术性现代主义将预示的某种绝对建筑，也没有必要在弗兰普顿[1]简称为"批判性地域主义"，即"区域学派"下来进行分类，这个主义的主要目的是要反映和服务于有限的、扎根的选区[2]。关于这些例子，最令人感兴趣的是他们的"地域性"并不一定意味着责任建筑师的出生地：这些都是全球性的建筑，与当地的条件相契合。

There is a wide gap between building and architecture, a kind of distance that the latter needs to cover in order to justify itself against the trivial. Most people mistake it for mannerism, projecting aesthetic quality as the foremost indicator of architecture's point of exclamation. Architects usually claim an interpretative stance against a given situation, the latter usually being a precedent in building form or program, or the contextual depository of the very land it occupies; it almost appears as if the principal requirement architecture has to meet is to elevate its practice away from the commonplace.

The cases we review here are set against a background. Anything but indifferent to their context, they propose the terms by which their defining elements relate to a situated condition. This is not the sort of overwhelming architecture your academic modernism would herald, nor does it necessarily classify under what Frampton[1] coins as "Critical Regionalism", namely "regional schools" whose primary aim is to reflect and serve the limited constituencies in which they are grounded[2]. What's most intriguing about these cases is that their "regionality" does not necessarily entail locality for the place of origin of the architects responsible: these are global architectures, in place with a local condition.

瓦山旅社_Washan Guesthouse/Amateur Architecture Studio
Brockholes访客中心_Brockholes Visitor Center/Adam Khan Architects
Beautour博物馆和生物多样性研究中心_Beautour Museum and Biodiversity Research Center
/Agence Guinée Et Potin Architectes
福古岛自然公园总部_Fogo Island Natural Park Headquarters/Oto Arquitectos
Sancaklar清真寺_Sancaklar Mosque/Emre Arolat Architects
玻璃农场_Glass Farm/MVRDV
do Morro竞技场_Arena do Morro/Herzog&de Meuron
水上的建筑_The Building on the Water/Álvaro Siza + Carlos Castanheira
在全球化建筑的背景下重塑建筑的地域性_Re-assessing Local Identity in the Context of Global Architecture
/Angelos Psilopoulos

Local Identity

我们所调查的项目与某些现有的条件相关。有趣的是，它们也可以被认为是彼此相关，事实上，当其他部分相对和谐地加入大群体中时，有一些也配对成功：

玻璃农场项目以一种对给定形式进行直接的重新解读的姿态来接受检查。该建筑为若干个单一的结构[3]，因其较为密集，因此与背景联系起来，且形成了明确的视觉表达。

Sancaklar清真寺呼应了福古岛国家公园的总部，它们都建立了一次与开放景观之间的对话，或分解为自然背景，或从自然背景的内部绽放。这两个项目都表现出一定的物质性，事实上，它们将其整体体验建此之上，直到其功能慢慢消失，分别衍变成确切的居住点。在这两种情况下，土地呈现本地化，因为它构成了实现这些建筑功能的原因。

最后，瓦山旅社、水上的建筑、Beautour博物馆和生物多样性研究中心、Brockholes访客中心有着共同的建筑术语，并主要作为景观来体验，起伏至它们所占据的地方，并给眼睛和身体提供了一座名副其实的"长廊建筑"。它们都既与周围环境有一丝关联，同时也与周围环境区分开来，它们看到外面的东西，并且俏皮地部署自己的感官体验。为了对固定的环境做出反应，这些建筑运用的材料和参考的概念都是为了发展其本身的建筑语言。然而，与此同时，它们也表明了它们不会冒然使用某些物质。

特别应该提到的是赫尔佐格&德梅隆建筑事务所设计的do Morro竞技场。在其形式品质内存在着明显的优点的同时，它还形成了一个更大的主题的一部分，即更广泛的、直接连接的、在城市规模内进行基础设施升级的一部分。从这个问题角度来看，它是一个在渴望愈合方面最开放的项目，我们没有专注于其本身的建筑语汇，而是提出检查其本身的诉求。

建筑

MVRDV设计的玻璃农场是为Schijndel的市场广场而开发的，是一处

All of the projects we examine relate to a certain existing condition. Interestingly enough, they can also be seen in relation to each other; in fact some pair up as others take part in larger groups quite congruously:

Glass Farm can be examined as a direct reinterpretation of a given form. It also forms "singular objects"[3] in relation to its backdrop as it is dense in itself, and makes a well-defined visual statement.

Sancaklar Mosque echoes the Fogo Island National Park Headquarters in establishing a dialogue with an open landscape, either by disintegrating into, or blooming from within, the natural backdrop. Both projects show a certain materiality; in fact they base their whole experience on it, to the point that their function fades into - or out to, respectively - the very land they inhabit. In both of these cases the land is the vernacular, in as much as it constitutes the very programmatic reason of the buildings.

Lastly, Washan Guesthouse, Building on the Water, Beautour Museum and Biodiversity Research Center and Brockholes Visitor Center share common terms in being primarily experienced as landscapes, undulating onto the premises they occupy and offering a veritable "promenade architecturale" to the eye as well as the body. They all both flirt with, as well as distinguish themselves from, their surroundings; they see what's out there and playfully deploy their own sensuous experience. In response to a situated context, these buildings employ materiality as well as conceptual reference to develop their architectural language. However, at the same time they show substance they do not offend.

Special mention should be made for Herzog & de Meuron's Arena do Morro. While there is an obvious merit in its formal qualities, it forms part of a larger scheme, namely, a much wider – yet directly connected – infrastructure upgrade on an urban scale. For that matter it is the one project most open about its healing aspirations. Rather than focusing on its own architectural vocabulary, we propose to examine it as such.

Architectures

MVRDV's Glass Farm was developed for Schijndel's market square, an "unusually void space in the fabric of the village". The 1600m² building shelters a series of public amenities such as restaurants, shops and a wellness Center. Years of debates with public officials and villagers alike, MVRDV's proposal covers the exact volume

Sancaklar清真寺，伊斯坦布尔，土耳其 Sancaklar Mosque, Istanbul, Turkey

"乡村结构内的不同寻常的上空空间"。面积为1600m²的大楼庇护了一系列的公共设施，如餐厅、商店和一个健身中心。就像多年与政府官员和村民的辩论一样，MVRDV的规划涵盖了城市规划者指定的确切体量和玻璃外围护结构。由于项目的体积等于传统的Schijndel农场的放大版本，因此MVRDV事务所和艺术家弗兰克·范·德·萨尔姆拍下这样的立面，通过加工直接作为拼贴画印制在玻璃上。此外，大楼的最终图像是一个合适的放大版，印制的透明度也有所不同，以允许光线和视野呈现多样化。在夜间，大厦的灯光产生了"传统农家乐"的感觉。与此同时，不成比例的元素带给游客一种不可思议的体验，类似于一个孩子遇到大人的世界。MVRDV最终实现了名副其实的Ceçi n'est pas une pipe（马格列的名画《这不是烟斗》）画面，一个不是农场的农场，是物质形态的虚拟表示形式，也是对其诠释开放的恒定辩证。

Emre Arolat建筑师事务所建造的Sancaklar清真寺是将其形状和体积融于景观的建筑，位于自然山体的下方，呈下沉状态，以展现洞穴般的室内。该建筑没有依赖于传统的清真寺采用的形式和特点，而是采用基本的氛围体验来传达其宏伟性。在祈祷大厅的里面，沿Qiblah墙面形成的缝隙和裂缝形成了一个戏剧性的照明效果，它能召唤精神的要求，以象征神的存在。从外面看，自然景观注入了不同规模的建筑元素：从定义石阶的斜坡到跨度为6m、形成天篷的钢筋混凝土薄板，再到石板墙和塔。这些元素能够对清真寺加以保护，并且人们在外面就能看见它。建筑师为我们提供了一些"奇怪的详细信息"，使陈腐的信息变得诗意化。就像建筑师说的："这个项目一直扮演着人为和自然之间直接的张力"，"只专注于宗教空间的本质"。

OTO建筑事务所设计的福古岛自然公园总部采取了相反的路线：建筑没有采用下沉景观，而是从内部部署。它坐落在"独特而罕见的美景"中，即天然的火山景观中，使用当地的技术和当地材料制成。由于其起伏的外形呈迷宫状，因此其所拥有的所有功能都采用集成且协调的方式，建筑既对"自然"提出了挑战，同时也将本土化提议转变为当代的规划。

designated by the city planners with a glass envelope. As this volume amounts to a scaled up version of a traditional Schijndel farm, MVRDV and artist Frank van der Salm photographed such a facade and imprinted it as collage directly onto the glass using frit procedure. The final image of the building is a "scaled-to-fit" enlargement of a regular one, the print also varying in translucency thus allowing for a variety of light and views. During the night the building illuminates, "offering tribute to the traditional farmhouse". At the same time, the out-of-scale elements bring the visitors an uncanny experience, similar to that of a child encountering the world of adults. Ultimately MVRDV achieves a veritable "ceçi n'est pas une pipe", a farm that is not a farm, a virtual representation in material form and a constant dialectic that is open to interpretation.

Emre Arolat Architects' Sancaklar Mosque dissolves its shape and volume into the landscape, sinking underneath the natural hill to reveal cave-like interiors. Rather than relying on the form and characteristics of a traditional Mosque it employs fundamental atmospheric experiences to convey the sublime. Inside the prayer hall, slits and fractures along the Qiblah wall provide for a dramatic lighting effect which calls upon the spiritual in order to signify the presence of God. From the outside, the natural landscape is infused with architectural elements of various scales: from the slope defining stone steps to the thin reinforced concrete slab spanning over 6 meters to form the canopy, to the stone clad walls and tower that protect the Mosque as well as reveal it from the outside, the architects offer us "strange details" that elevate the trite to the level of the poetic. As the architects say, "the project constantly plays off of the tension between man-made and natural", "focusing solely on the essence of religious space".

Oto Arquitectos' Fogo Island National Park Headquarters take the opposite route: rather than sinking into the landscape, the building appears to deploy from within it. Situated in a natural volcanic landscape of "unique and rare beauty", the building is made from local materials using local techniques. As it undulates into a maze-like shape that nests all functions in an integrated and harmonious way, the building both challenges the "natural" as well as transforms the vernacular into a contemporary proposal. Taking the architects' claim at face value, "with the headquarters fully operational, the natural park is increasingly valued which contributes

照片提供：©Amateur Architecture Studio

瓦山旅社，杭州，中国 Washan Guesthouse, Hangzhou, China

从表面价值来看，用建筑师的说法，就是"总部全面运作起来的同时，自然公园也越来越受重视，这有助于丰富岛上的社会、文化和经济领域，以和谐的方式开始整合和提升周围空间"。

王澍的瓦山旅社是专为中国美术学院设计的，作为校园十多年不断发展的建筑群的一部分，这些建筑的主要目的在于教学，且在他的建筑和追随文人传统的脚步方面起到了积极的作用。坐落在杭州，一座古老的城市，因自然和文化遗产而闻名，旅社以富有想象力的建筑课程的形式，展现了传统的主题，重新诠释了中心舞台。本地化仅作为微不足道的参考：一方面，本地化存在于建筑的物质性中，如使用当地乡土结构中的夯土。另一方面，建筑设有100m长的屋顶，起伏的景观位于其上，层叠的斜坡平面将景观"框"入其中，因本身也作为景观在不断地变化着。这是一个对山体景观的直接类比，文人常用此对比来进行绘画，以传达一连串的经历或感受（而非单独的画面）。最后，一系列不规则的木支柱推动了共同环境之外的传统形式和技术。这些行为都转译为诗歌，可能是瓦山旅社所必须提供的最显著的教导。用王澍的话说，感受和经历无法做成一首诗，除非它们被安置在某种结构中。

阿尔瓦罗·西扎和Carlos Castanheira的"水上的建筑"以一种不同的当地性来与我们交谈，即一个工业建筑群的技术效率。位于江苏省淮安市新盐工业园区内，按照公司负责人的期望，该建筑建在一个现存的、服务于工厂的人工湖上。它以曲线的形式进行延伸，邀请我们进入浮动的、身体几乎不受控制的外在体验中，而工业技术网格仅作为这条"龙"的栖息地来接待来访者和用户。围绕建筑师们的决定，时间暂停，空间扭曲：在这里，这种做法融合了雕塑的做法，"因为它曾经是一个雕塑"。"水上的建筑"是指通过建筑揭示水的诗性来展示其工业（即功能）遗产。这与它假定其身份和分配价值是原本平庸的背景的方式相同。

Agence Guinée Et Potin的Beautour博物馆和生物多样性研究中心遵循了前两个项目的脚步。该项目恢复了法国博物学家乔治·杜兰德府邸的"荣耀"，"旨在提供有关生物多样性的教育和科学支持，以及整个地

to enrich the social, cultural and economic sectors of the island, starting to integrate and enhance in a harmonious way the surrounding space".

Wang Shu's Washan Guesthouse was designed for the China Academy of Arts as part of an ever evolving cluster of buildings he has been adding to its campus for over a decade. Taking an activist role in his architecture and following in the footsteps of the literati tradition, these buildings are primarily aimed to teach. Situated at Hangzhou, an ancient city renowned for its natural and cultural heritage, the guesthouse brings the themes of tradition and reinterpretation of center stage in an imaginative lesson of architecture. The vernacular exists in subtle reference: for one, it lies in the materiality of the building, like the use of rammed earth found in local vernacular structures. And then there is the undulating landscape of a 100m long roof, a superimposing cascade of sloping planes that frame the landscape in as much as they evolve as landscape in themselves. This is a direct analogy to the mountainscapes the literati used in painting in order to convey – rather than a single image – a series of experiences or feelings. Finally, an irregular nebula of wooden struts pushes the traditional forms and techniques outside their common context. These acts of translation are revealed as poetry, probably one of the most significant teachings the Washan guesthouse has to offer. In the words of Wang Shu, feelings and experiences do not make a poem until they have been placed in some kind of structure.

Álvaro Siza and Carlos Castanheira's "Building on the Water" converses with a different kind of locality, namely the technical efficiency of an industrial complex. Situated within the premises of the New Salt Industrial Park of Huai'an City, Jiangsu Province, the building is built on top of an already existing artificial lake that serves the factory, by the desire of the company's owner. It extends in a curvilinear form, inviting us into a floating, almost out-of-body experience, the technical grid of industry serving as mere ground for this perched dragon to receive visitors and users alike. Time suspends and space is warped around the architects' decisions: this practice here is fused with the practice of sculpture, "as it used to be". The "Building on the Water" refers to its industrial (namely functional) heritage by revealing the poetic potential of the latter through architecture. This is the very same way to assume its identity and assign value to a trite backdrop.

do Morro竞技场，纳塔尔，巴西 Arena do Morro, Natal, Brazil

区的管理策略和发展前景"。围绕周围的现有建筑，该中心完全以传统工艺，即茅草皮，来覆盖它的外壳，一种全新的技艺被重新诠释成一个新的想法，如"一片建筑景观""一个新地形"，形成了自然的透视景，有机、有动物，为背景做出贡献。除了作为一个教育和科研机构，它的目的还在于作为维护现有场地（已经废弃了30年）的再生力，我们再一次发现在指示的领域内，建筑承担了恢复张力的孵化器的角色。

Adam Khan建筑师事务所的Brockholes访客中心作为一组空间，漂浮在"大浮桥"之上。该建筑融入到67公顷的自然保护区景观中，以证明其设计目的，并作为建筑的标志高高耸立。"建筑"以抽象的形式存在：鹅卵石屋顶以不寻常的高度和体积呈几何形状倾斜着，反而形成内部空间氛围。墙壁上覆盖着芦苇，屋顶覆盖着橡木面板，整座建筑显得十分"自然"，但并非如此。在建筑物内，屋顶极致的几何形状由技术要求很高的"编织"木框架支撑，然而表面呈现的微妙触感让它们完全有吸引力邀请游客进入。广泛使用的大型玻璃窗格使屋顶和地面分离，然而它们却不是"顶点"，因为它们的脊部被天窗所取代，天窗将外部光线反射回室内空间。建筑无需采用本地的建筑语言，而是通过非物质性来向我们展示其自身，并且产生了"不可思议的熟悉度"。这是一种定义了整个体验的矛盾性原理。

赫尔佐格&德梅隆建筑事务所设计的do Morro竞技场属于一个"按时间来建造的嵌入结构"规划的一部分，也是一个传达公司的"麦·路易莎地区视野"理念的基础补充设施——当地官方希望通过建筑来支持的地区和社区。建筑本身作为一个体育、文化和社会中心，它的形式演变为一个全保护屋檐下起伏的沙丘。屋顶尽可能多地从远方覆盖，并且在地平面的视野内消失：它分解于天空，似乎它由重新定义的材料，而非纯粹的技术肌理制成。建筑不仅仅是作为一个标志，而且还包括了一处大型开放空间，用于社区集会。在同一时间，起伏的城墙为较小规模的活动提供了庇护的山凹。这些外观质量应该与一系列规划的建筑一起进行检查，作为更大计划的一部分。它允许以集成的方法来处理这个问

Agence Guinée Et Potin's Beautour Museum and Biodiversity Research Center follows the footsteps of the previous two projects. "Glorifying" French naturalist, Georges Durand's mansion, the project "aims to offer educational and scientific support on the theme of biodiversity, as well as a management strategy and an evolution prospective for the whole area". Revolving around the existing building, the center covers its shell fully in thatched skin – a traditional technique reinterpreted into a novel idea – lying thus as "a piece of built landscape", "a new geography" completing the natural scenography, organic and almost with animals, in tribute to its context. Besides serving as an educational and a scientific facility, it also aims to act as a re-generative force for the maintenance of the existing site, which has been abandoned for 30 years. Once again we find ourselves in a signifier territory, architecture assuming the role of an incubator of reinvigorating tensions.

Adam Khan Architects' Brockholes Visitor Center spans as a cluster of spaces floating on top of "a large pontoon". Being integrated into the natural landscape of a 67-hectare nature reserve to justify its very own programmatic purpose, it stands primarily as an architectural signifier. The "houses" take their forms through abstraction; the shingled roof skews its geometry to an unusual height and volume, shaping in turn the atmosphere of the interior space. Walls are clad in reeds, roofs are covered with oak shingles, and the whole thing appears "natural" and yet not. At the inside of the building the extreme geometry of the roofs is supported by a technically demanding "woven" wooden frame – yet the surfaces assume a subtle tactile feeling which utterly makes them inviting. The roofs detach from the ground by the extensive use of large window panes - yet they never "culminate" as their ridge is substituted by a skylight that throws exterior light back into the interior space. Rather than a narrative taking on the vernacular, architecture offers us presence through immateriality, in conjunction with an "uncanny familiarity" – a contradiction that defines the whole experience.

Herzog & de Meuron's Arena do Morro is attached to a larger scheme of "punctual interventions", an infrastructure supplement conveying the firm's "Vision for Mãe Luiza" – a land as well as a community which the local authorities hope to support through architecture. The building itself serves as a sports, cultural and social center, its form evolving as dune-like undulations sheltered under an all-protecting roof. As much as it is imposing from afar,

1. Kenneth Frampton, *Modern Architecture: A Critical History(World of Art)*, 3rd ed., rev. and enlarged, London: Thames and Hudson, 1992.
2. Ibidem, p. 314.
3. Jean Baudrillard, *The Singular Objects of Architecture*, Minneapolis: University of Minnesota Press, 2002.
4. Bart Verschaffel, *Architecture Is (as) a Gesture*, Luzern: Quart Verlag, 2001, p. 29.
5. Ibidem.

题：该项目更多地被归类于基础设施和城市规划，较少地关注其形式、叙述方式和风格。在此背景下，建筑通过福音的方式从一个“痛苦”的背景下区分开来，来揭示其愈合的愿望。

本地身份和宣称的卓越性

绕过建筑无所不在的增添价值的宣言，我们所有的案例似乎都引发了一个争议，即“全球性”的建筑是否能够成为它占用的地面的一部分。一方面，其本身很难呈现本地地形。在所有的情况下，设计师重新诠释了当地的条件，赋予新的特质，而最重要的是，提出了一种新的理解，包括什么是“当地问题”。这似乎是一个专注于当代建筑的当务之急：从常见的形式到对其环境的全新理解方面来提升其规划。

在这些条件下，建筑自称在其所占据的地面中应用某一种增值来证明自己，无论是在概念还是在实际的条款下。与此同时，人们仍然需要付出代价：本地化永远不能像以前一样，人们永远无法继续生活在共同地方的安全怀抱里。这种建筑不再是木匠和石匠建成的“建筑”，也不与它号称的景观一样“天然”。这预示着，首先，也是最重要的是，它是一个出发点而不是普通形式的“故障性”重复。

按照这样的说法，建筑是完全现代化的。我们研究的项目以它们最温和的形式提出了辩证；以他们最大胆的方式重新解释了重塑理念性的程度。作为价值衡量，这只能由天才[4]加以判断，而不是由建筑师本身判断，否则只会出现一个“高度膨胀的自我”[5]，这将现有的局面转变为智者定期实现的壮举。

the roof dissolves at ground level view: it disintegrates into the sky, seeming like it's made from re-purposed materials rather than a purist techno-fabric. More than a signifier, it covers a wide open space for community gatherings, at the same time the undulating walls offer sheltering coves for smaller scale activities. These formal qualities should be examined in conjunction with a sequence of proposed buildings – as a part of the larger scheme of things. This allows for an integrated approach to the problem at hand: the project is more infrastructure and city planning and less form, narrative and style on its own. In this context, architecture reveals its healing aspirations by distinguishing itself from a "suffering" backdrop in an almost evangelical manner.

Local Identity and the Claim to Excellence

Bypassing architecture's omnipresent claim to add value, all of our case studies seem attached to an argument whether a situated "global" architecture can form part of the ground it occupies. For one thing, it is hardly vernacular in itself. In all of our cases, the architects re-interpreted the local condition, assigned new qualities, and most of all, proposed a new understanding for what constitutes "a local problem". This seems to be a constant preoccupation for contemporary architecture: to elevate its proposal from a common form to a novel understanding of its context.

In these terms, architecture justifies itself by claiming to a certain kind of added value it applies to the ground it occupies, be it in actual terms or conceptual ones. At the same time, a price is to be paid: the vernacular can never be the same as before, and people can never continue their lives safely in the reassuring arms of the commonplace. This architecture is hardly the carpenters' and the stone masons' building; nor is it as "natural" as the landscape it lays claim to. It heralds, first and foremost, a point of departure rather than a "failsafe" repetition of the common form.

By that argument, this architecture is utterly modern. The projects we examined offer, in their most modest, a dialectic; in their boldest, they offer a full reinterpretation to the extent of conceptual rebirth. As a measure of value, this can only be justified by a claim to genius[4] – yet not by an exclamation of the architect's own, whence only a "highly inflated ego"[5] would appear, but by the transformation of an existing situation into a regular feat of the intellect. Angelos Psilopoulos

瓦山旅社

Amateur Architecture Studio

这个项目由王澍和他的杭州团队所设计，位于杭州的象山园区。这是一个生活辅助设施，可以满足日益增长的游客食宿的需要。场地位于象山山峰南侧，沿河而建，南北窄，东西长。建筑面积超过5000m²。但对于这个场地来说，建筑的体积可能太大，高度也可能太高。这就很难处理建筑与山、树木和河流之间的关系。

从建筑师的角度来看，场地内的建筑形成特殊的氛围是最重要的。这里的大自然充满了潜在的诗意。优秀的建筑能够揭示出与环境氛围之间的沉默对话的这种诗意。看似最简单的解决办法是在象山山峰南侧建造一座带有层次的小山。但这是一个满山覆盖着回收的旧瓦片的小山，命名为"瓦山"。

这个设想也来自于建筑师二十年前在湖南省西部山区旅行的经历。一个村庄长久以来沿河建造。这里有数百座带有斜屋顶的折叠式传统民居，由一条蜿蜒的瓦片屋顶相连接，街道和小巷都覆盖着一个巨大且连续的瓦屋顶。这个巧妙的、解决雨季和炎热夏天气候的办法也隐含了想象和诗意。

在建筑师最新设计的120m长的瓦屋顶下，松木条形成了大跨度的空间。下面是三十多面60cm厚的夯土墙，将建筑分为六个独立的单元。从东向西分别是茶馆、会议中心、餐厅、三个庭院模式的酒店。考虑到抗震规则的严格要求，混凝土框架结构用来支持夯土墙。

从东西方向看，建筑看上去像一座小山，呈曲折状态。从南北方向看，建筑看上去像一座小山状的、通风的屏风；人们可以通过建筑物看到室外，这样感觉建筑物的体积从视觉上减小了。这一结构就像一个小的建筑群笼罩在一个巨大的棚里。双屋顶不仅抵挡了夏季的炎热，也形

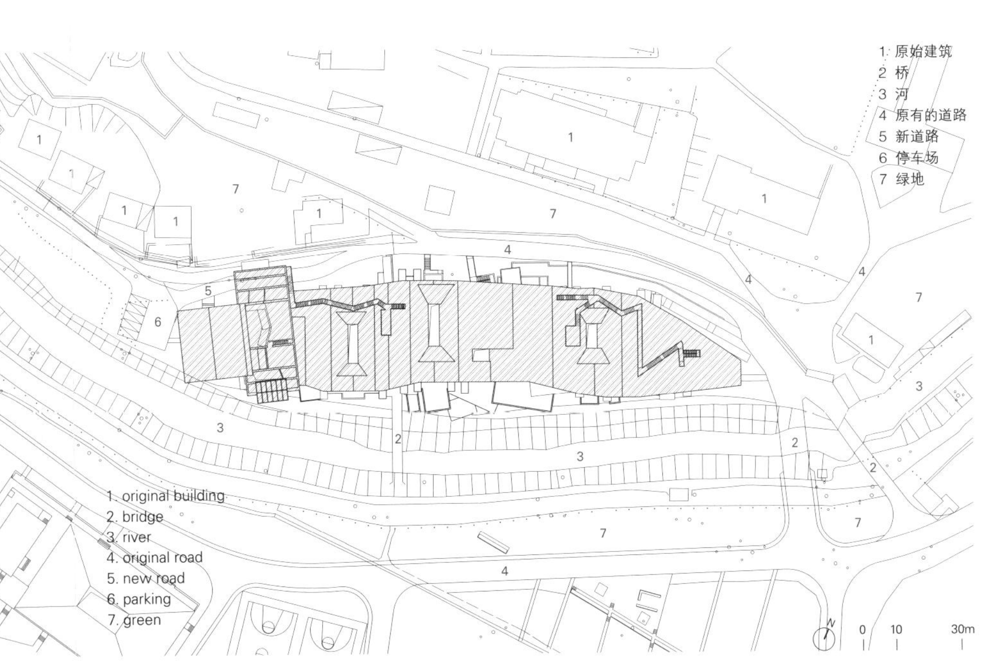

1. 原始建筑
2. 桥
3. 河
4. 原有的道路
5. 新道路
6. 停车场
7. 绿地

1. original building
2. bridge
3. river
4. original road
5. new road
6. parking
7. green

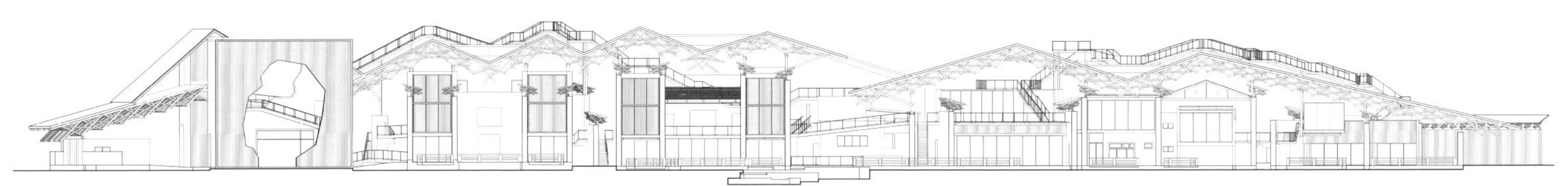

南立面 south elevation

北立面 north elevation

成了令人印象深刻的空间。在多雨的季节，人们可以随意地生活在棚下，自由地走动。

有三种方式可以观察和体验这座山。沿着河流，有一条路径穿过建筑，并穿过许多夯土墙上的大门。

层层叠叠，许多平台垂向河中，类似建筑师在京都鸭川河边旅行所经历的那种风格。在建筑的中间部分，还有另外一条路贯穿整座建筑，一路上起起落落。第三条道路是在屋顶上，蜿蜒前行。从远方望去，感觉像是建筑师最喜欢的五代时期画家董源的画。

Washan Guesthouse

This project is located on the Xiangshan Campus designed by Wang Shu and his team in Hangzhou. It is a supplementary facility of living also to meet the increasing needs of the visitors' accommodation. The site is on the south side of the Xiangshan, along the river, narrow from the south to north and long from the east to west. The building area is about 5,000 square meters more. But to this site, the volume of the building might be too big and the height might be too high. It is difficult to handle the relationship between the building and the hill, trees and the river.

From the architect's point of view, the special atmosphere formed by a building on a site is the most important. Nature here is full of potential poetic. A good building can reveal this poetry in the silent conversation with the environmental atmosphere. The seemingly simplest solution is to build another layer hill on the south side of the Xiangshan. But this is a hill covered with recycled old tiles, named "Tiles Hill".

This assumption was also derived from the architect's travel experience in the west mountain area of Hunan Province twenty years ago. A village had been constructed along the river. There were hundreds of traditional fold residential houses with sloping roofs, connected by a winding tile cover and the streets and lanes were all covered with a huge continuous tile roof. This smart solution to the rainy and extremely hot summer climate also implied imagination and poetry.

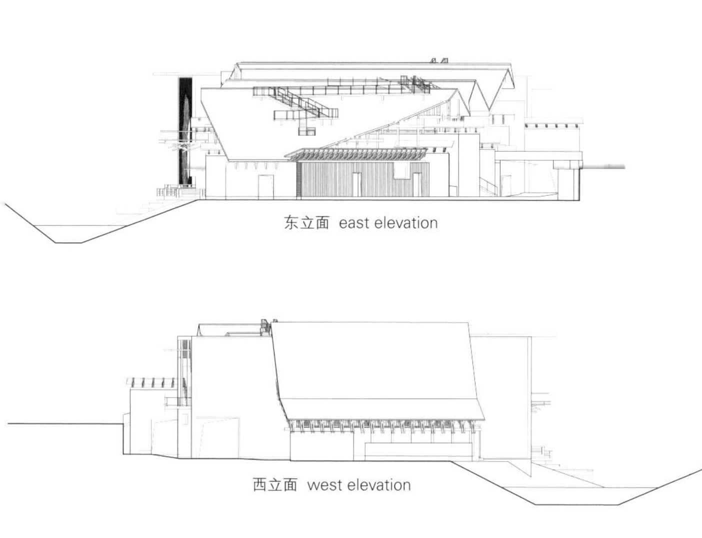

东立面 east elevation

西立面 west elevation

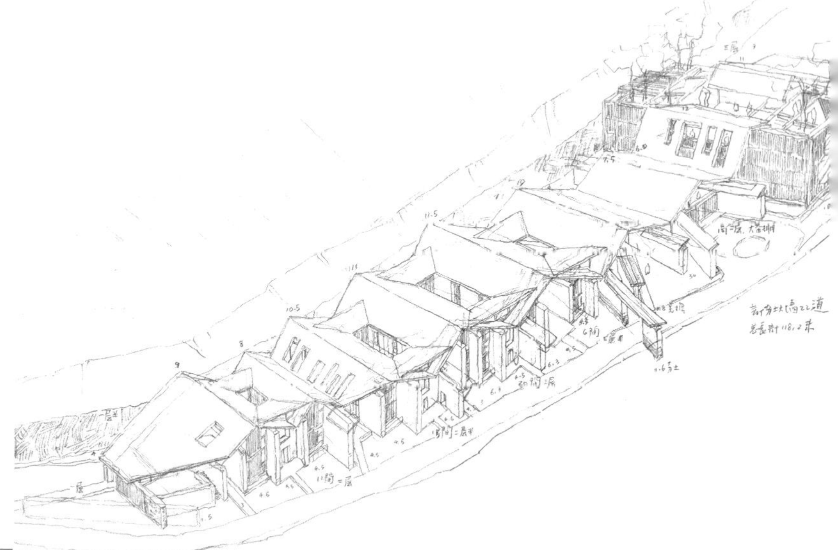

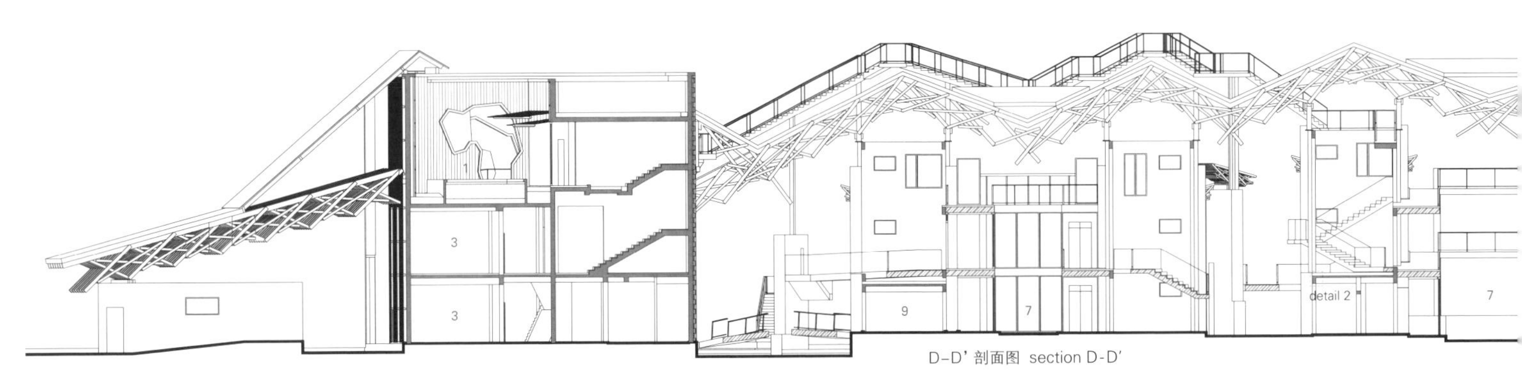

D–D' 剖面图 section D-D'

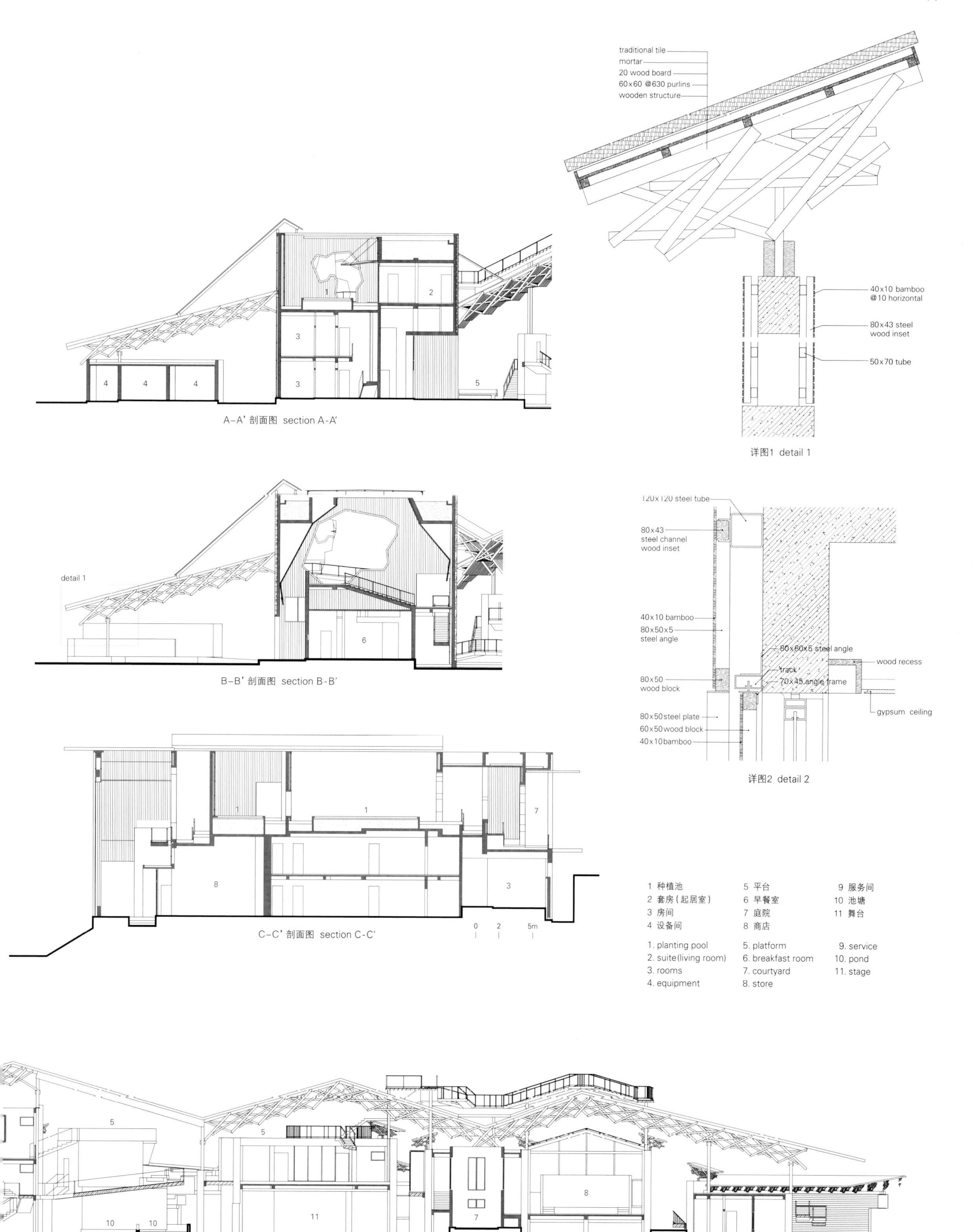

A–A' 剖面图 section A-A'

详图1 detail 1

B–B' 剖面图 section B-B'

详图2 detail 2

C–C' 剖面图 section C-C'

1 种植池
2 套房(起居室)
3 房间
4 设备间
5 平台
6 早餐室
7 庭院
8 商店
9 服务间
10 池塘
11 舞台

1. planting pool
2. suite(living room)
3. rooms
4. equipment
5. platform
6. breakfast room
7. courtyard
8. store
9. service
10. pond
11. stage

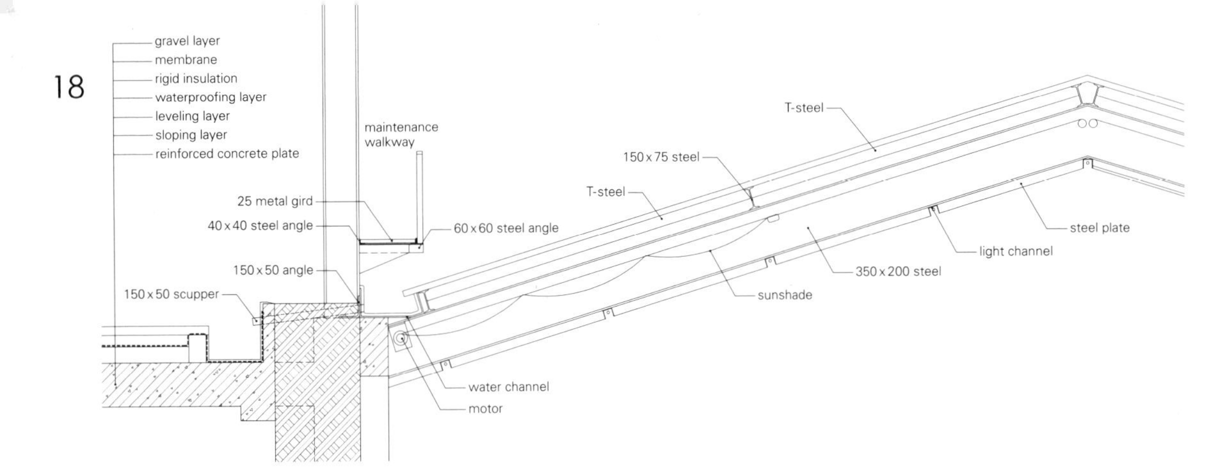

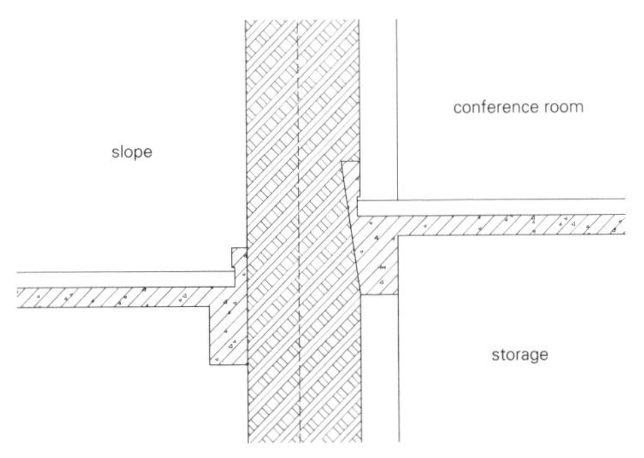

a–a' 剖面图 section a-a'

项目名称：Tiles Hill - New Reception Center for the Xiangshan Campus / 地点：Zhuantang, Hangzhou, China
建筑师：Wang Shu, Lu Wenyu
设计团队：Chen Lichao / 结构工程师：Shentu Tuanbing / 设备工程师：Sun Mingliang
斜坡场地的技术支持：Marc Auzet, Juliette, Zhang Wen, Wei Chaochao / 木框架设计助理：Jeremiah Watson
甲方：China Academy of Art
用地面积：7,500m² / 总建筑面积：2,596m² / 有效楼层面积：around 6,200m²
结构：concrete with steel bars frame, rammed earth wall, wooden structure
材料：bamboo plank concrete, recycled tiles, recycled ceramic pieces, earth, timber
设计时间：2010.6—2013.5 / 施工时间：2011.8—2013.8 / 造价：USD 9,100,000
摄影师：
©Wang Junlei(courtesy of the architect)-p.20top / ©Wang Tiantian(courtesy of the architect)-p.10~11, p.20bottom
©Edward Denison-p.12~13, p.14~15, p.16, p.18, p.19, p.21, p.23, p.24, p.25, p.26~27

屋顶 roof

1 设备间 2 种植池 1. equipment 2. planting pool
屋顶_木框架结构 roof_wooden frame structure

1 庭院 2 种植池 3 套房(起居室) 4 套房(卧室) 5 会议室 6 房间 7 平台
1. courtyard 2. planting pool 3. suite(living room) 4. suite(bedroom) 5. conference 6. rooms 7. platform
三层 third floor

1 房间 2 办公室 3 服务间 4 管理间 5 商店 6 池塘 7 舞台 8 平台 9 会议室 10 讲厅 11 接待处 12 茶室
1. rooms 2. office 3. service 4. management 5. store 6. pond 7. stage 8. platform 9. conference 10. lecture hall 11. reception 12. tea room
二层 second floor

1 入口大厅 2 大堂 3 接待处 4 早餐室 5 办公室 6 西餐厨房 7 设备间 8 房间 9 平台 10 池塘 11 庭院 12 服务间 13 餐厅 14 舞台 15 中餐厨房 16 餐具室 17 礼堂 18 管理间 19 商店 20 茶室 21 会议室 22 讲厅 23 套房(起居室) 24 套房(卧室) 25 种植池
1. entrance hall 2. lobby 3. reception 4. breakfast room 5. office 6. western kitchen 7. equipment 8. rooms 9. platform 10. pond 11. courtyard 12. service 13. restaurant 14. stage 15. Chinese kitchen 16. pantry 17. hall 18. management 19. store 20. tea room 21. conference 22. lecture hall 23. suite (living room) 24. suite (bedroom) 25. planting pool
一层 first floor

Under the 120-meter-long tile roof newly designed by the architect, the pine wood bars form a space of large span. Below it are more than thirty rammed earth walls of 60cm thick, dividing the building into six independent units. From the east to west are respectively the tea house, conference center, restaurant, three courtyard pattern hotels. Taking into consideration of the strict earthquake-resistance rules, concrete frame structure is used to support the rammed earth walls.

Viewed from the east and west, the building looks like a hill, twists and turns. Viewed from the south and north, the building looks like a hill-shaped ventilating screen; people could see through the building so that the volume of the building reduces visually. The structure is like a small group of buildings shadowed by a huge shed. The double roof not only keeps out the heat in summer but also emerges as an impressive space. In rainy seasons, people could live under the shed casually and walk around freely.

There are three ways to observe and experience this hill. Along the river, there is a path passing through the building. It passes many gates on the rammed earth walls.

Layer upon layer, many platforms hang towards the river, sort of similar as the river side of Kamo River in Kyoto derived also from the architect's travel experience. In the middle part of the building, there is another path through the whole building, rising and falling on the way. The third path is on the roof, climbing and winding. Looked at the distance, it feels like the architect's favorite Wudai Dynasty painter Dong Yuan's painting.

Brockholes访客中心

Adam Khan Architects

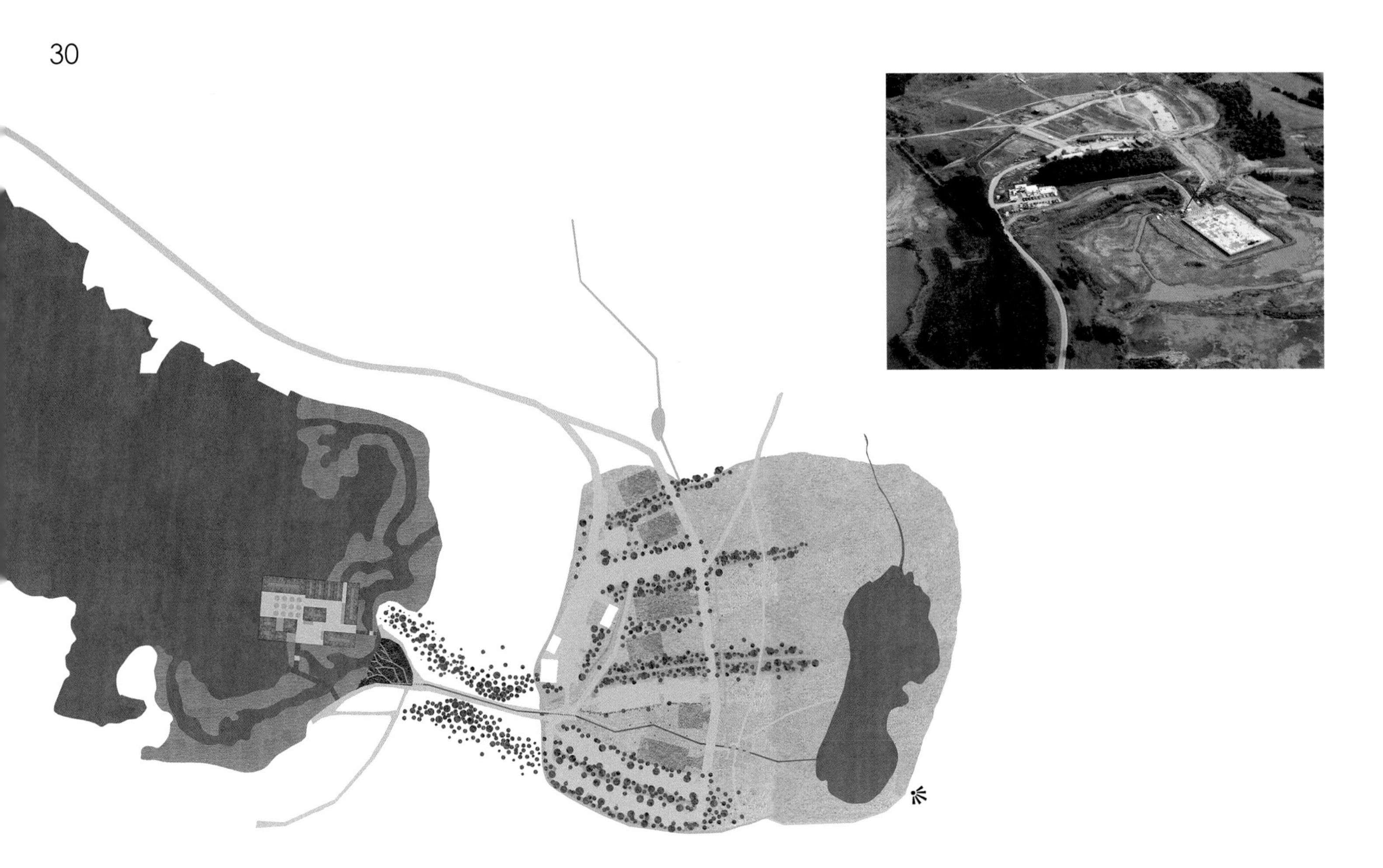

客户对这一项目的野心是有远见且深刻的。作为一个经验丰富的野生动植物信托基金会，其成员清楚地看到激励大范围的人们去爱和享受自然，且超越专业人士的需要。这既是全球的也是当地的需要。这个星球所面对的大环境的挑战将需要明智且热情的支持者，而基金会是这一活动的重要角色。但是和在国家性和战略性层面工作一样，他们的工作牢牢植根于当地的法案。普雷斯顿拥有一些惊人的自然景点，但同时场地附近也拥有被严重剥夺自然资源的口袋。之后项目书要求在这些层次上进行工作，并且创造一处鼓舞人心的、高度可持续发展的自然保护区。

关于这一开放式项目书的RIBA竞赛举办于1997年，Adam Khan建筑师事务所的方案"漂浮的世界"赢得胜利，该事务所领导了保护区的整体规划，开发了游客中心和相关的景观工程，随着业务规划的发展，反复规划工作以确保实际且经济的可持续保护区。

仓库所展现的木结构具有创新性，连同工程效率作为一个更为深刻的目的——"结构的感觉"的基础。覆层面板创新性地作为应力蒙皮来使用，允许框架变得更薄，形成一个连续的、具有精确的垂直度的网络结构。允许这个系统适应不对称性，能产生微妙的活力和空间的连续性，这种连续性掩盖了工程和制造要求它所达到的复杂性。预制胶合木框架结构提供了低重量、低碳且精确的切割，复杂的几何形状，以及快速安装和统一质量的服务。SIP板形成一个高度保温的轻型屋顶，无热桥。窗户沿用一个独立的模块，减轻了外围护结构构件的质量，且对其进行了分层。

声波喷雾的触感和深颜色使其结构外壳变得抽象，且赋予了草仓库毫无生气的寂静感，以唤起人们对房间和庇护所的触觉反应。高高的仓库参考了集体空间原型的记忆，并制定了一套适用不同时间尺度来居住的基础设施。高高的屋顶，具有形式的重复性和鲜明的椽木裂缝，位于一个更加协调的人类活动层上，复杂的自然通风设备在此集成。未经处理的椽木覆层与无框玻璃、ocalux屏风以及穿孔的黄铜构件一起完成，从而赋予不平的、光亮的、复杂的水反射面一个镶花表面。一个低矮的屋檐使密集的屋顶触手可得，且匹配了成熟的芦苇层，芦苇层环绕着浮桥——屋顶将从芦苇中伸出，中间的空间就像麦田圈一样。本地采购的骨料作为地板表面显露出来。

Brockholes Visitor Center

The client's ambition for the project is visionary and profound. As an experienced Wildlife Trust, they see clearly the need to inspire a broad range of people to love and enjoy nature, and reach beyond the specialist audience. This is both global and local. The big environmental challenges facing the planet will need an informed and passionate constituency, and the trust is a key player in this. But as well as at national and strategic level their work is firmly rooted in local action. Preston enjoys some stunning natural sites but also pockets of severe deprivation lying close by the site. The brief then was to work at all these levels and create an inspiring and highly sustainable nature reserve.

A RIBA competition with this open brief was held in 1997, won by Adam Khan Architects with the scheme "A Floating World", who then led the masterplanning of the reserve, and developed the visitor centre and associated landscape works, working reiteratively with the development of the business plan to ensure a

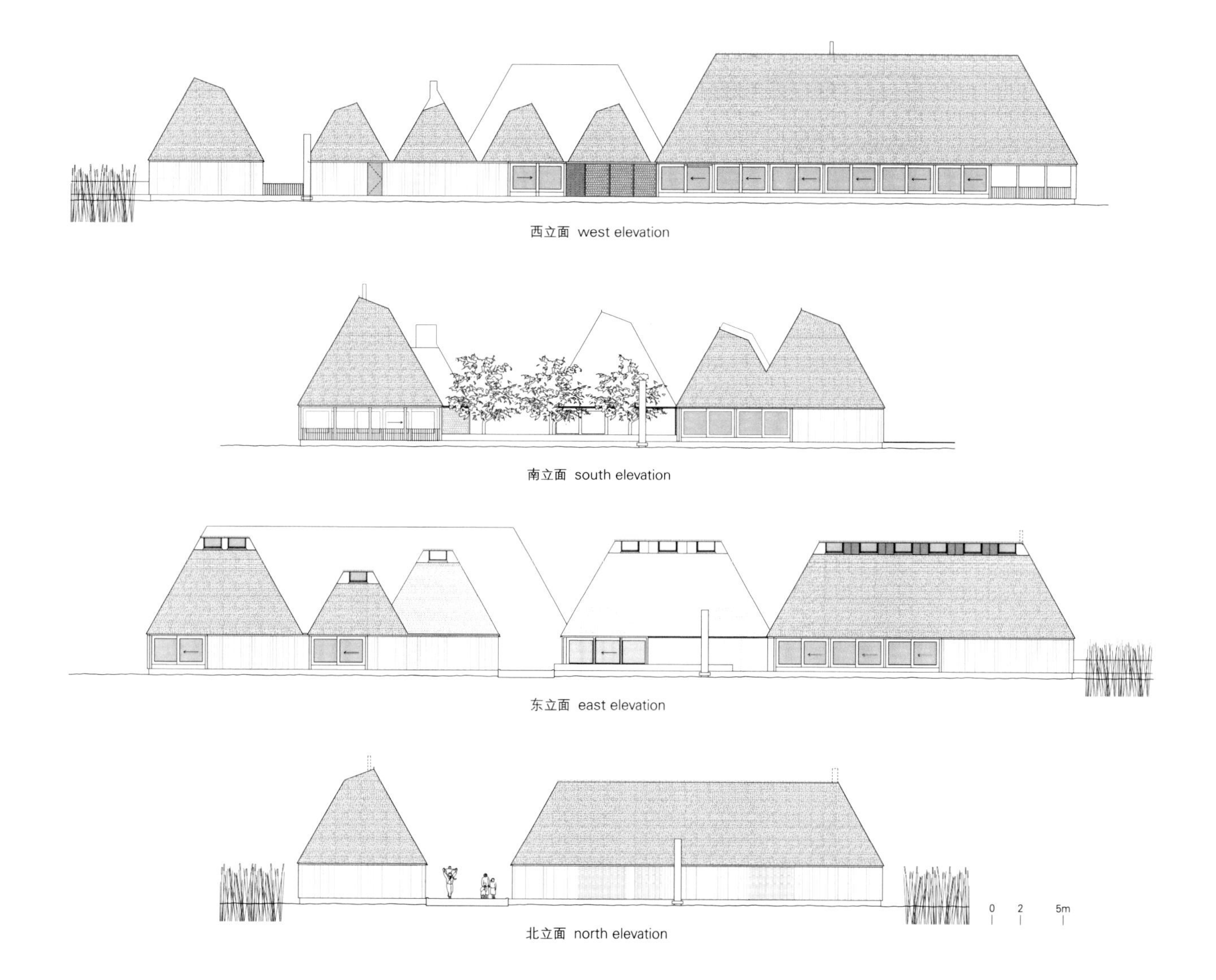

西立面 west elevation

南立面 south elevation

东立面 east elevation

北立面 north elevation

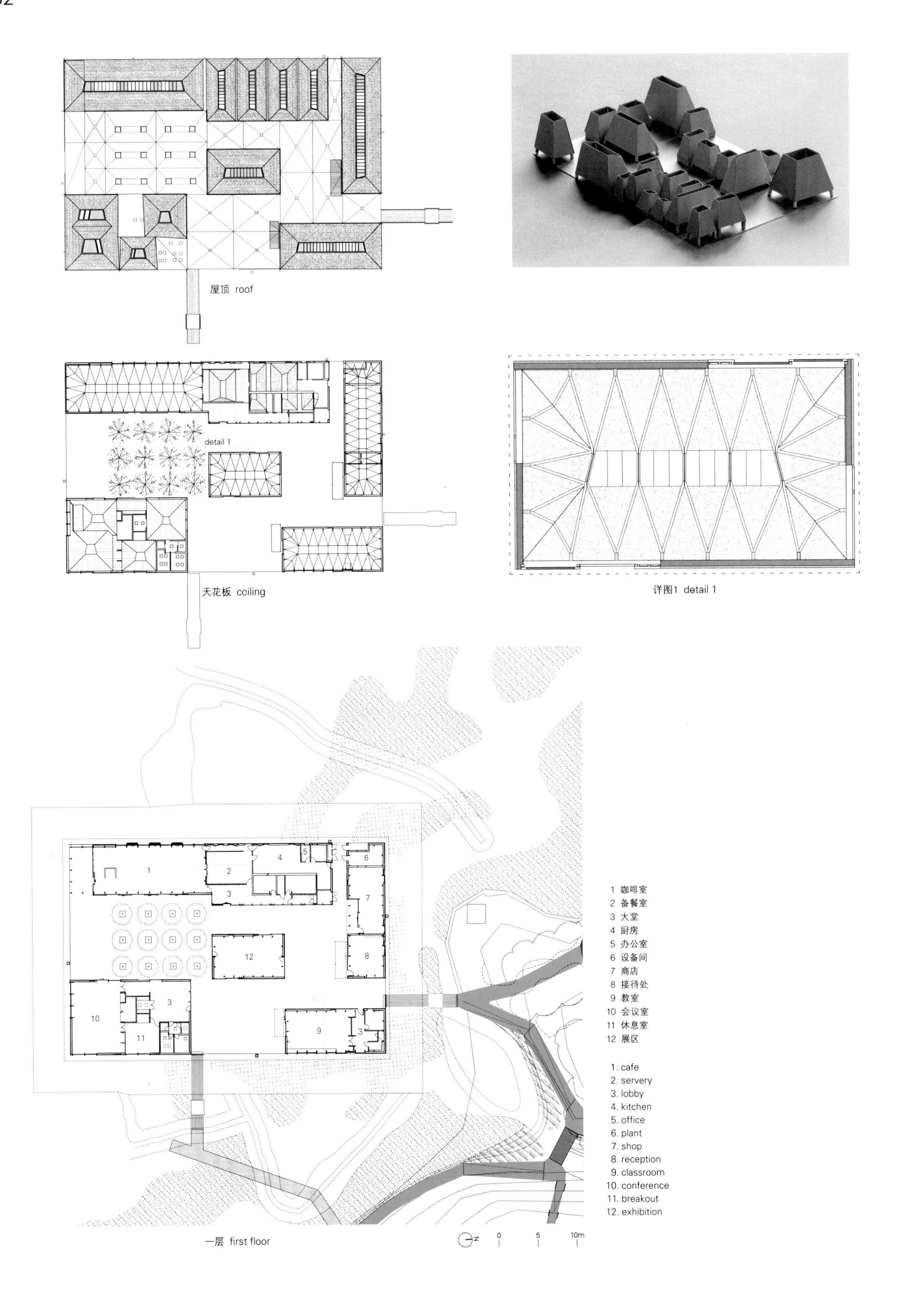
屋顶 roof
detail 1
天花板 ceiling
详图1 detail 1
1
2
3
4
5
6
7
8
9
10
11
12
一层 first floor
0
5
10m
1 咖啡室
2 备餐室
3 大堂
4 厨房
5 办公室
6 设备间
7 商店
8 接待处
9 教室
10 会议室
11 休息室
12 展区
1. cafe
2. servery
3. lobby
4. kitchen
5. office
6. plant
7. shop
8. reception
9. classroom
10. conference
11. breakout
12. exhibition

practical and financially sustainable reserve.

The expressed timber structure of the barns uses innovation and engineering efficiency as the basis for a more profound intention– the "feeling of structure". The innovative use of cladding panels as stressed skin allows the frame to be thinned out, exploited to create a continuous net of structure with a refined verticality. Allowing this system to adjust to the asymmetry gives a subtle vitality and spatial continuity which belies the complexity of the engineering and fabrication required to achieve it. The prefabricated glulam framed structure offers low weight, low carbon, accurate cutting and achievement of complex geometry, rapid erection and consistent quality. The SIP panels form a highly insulating, lightweight roof, free of thermal bridging. The windows follow an independent module from the structure, lightening and layering the components of the envelope.

The tactility and deep color of the acoustic spray abstract the structural shells and give the acoustic deadness of a hay barn, evoking a haptic response of nest and shelter. The lofty barns reference an archetypal memory of collective spaces, and make a set of adaptable infrastructures for inhabitation on differing timescales. The tall roofs, with their repetition of form and vivid materiality of oak shakes sit on a more tuned human layer, where sophisticated natural ventilation is seamlessly integrated. Cladding of untreated oak is finished flush with frameless glazing, ocalux screens and perforated brass, giving a marquetry surface of rough and polished, and complex watery reflections. A low eaves brings the dense mass of roof within reach, and matches the datum of the mature reeds which will surround the pontoon – the roofs will emerge from the sea of reeds, the spaces between carved like crop circles. Locally sourced aggregate is exposed as floor surface.

nce Centre

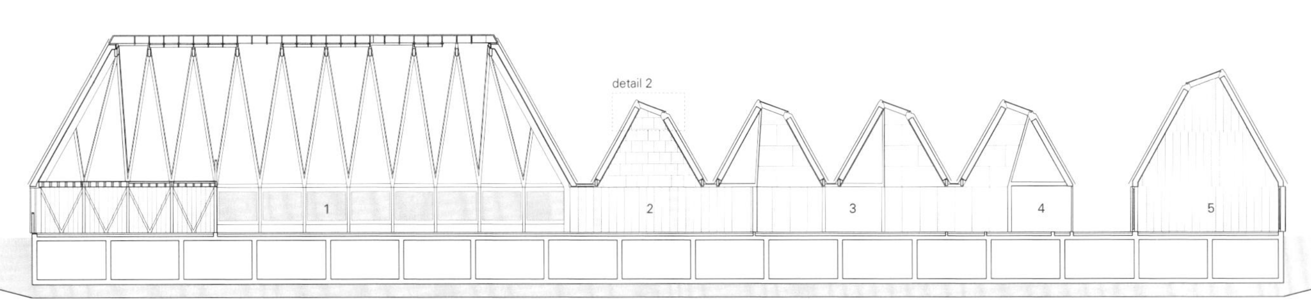

1 咖啡室 2 备餐室 3 厨房 4 办公室 5 设备间 1. cafe 2. servery 3. kitchen 4. office 5. plant
A-A' 剖面图 section A-A'

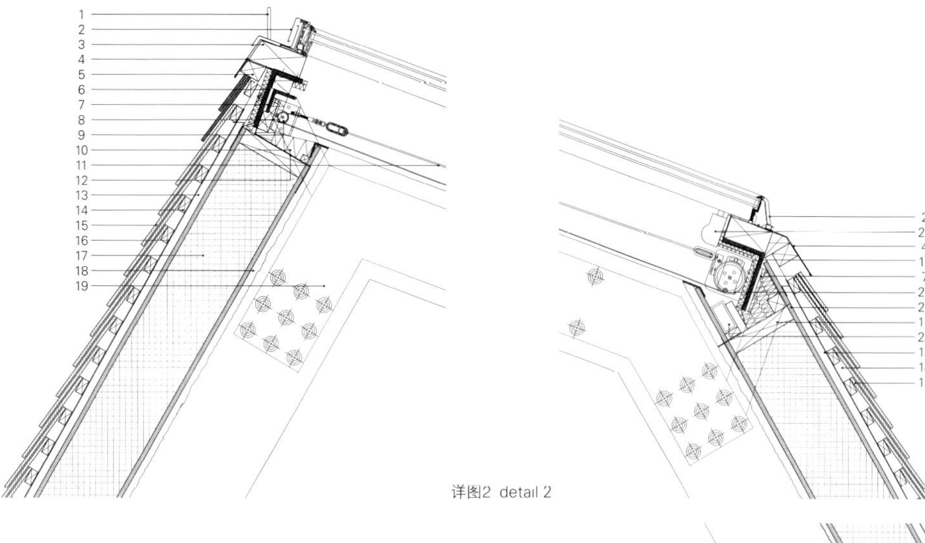

详图2 detail 2

1. bird deterrent
2. rooflight flashing (L10C/460)
3. treated SW plate
4. copper flashing(H73/100)
5. intermittent packing spacers
6. ridge frame as per S.E. spec
7. thermal break
8. blind head plate
9. timber packer shaped to suit
10. bracket as per S.E. spec
11. blind guide wire
12. air tightness membrane
13. counter batten
14. tiling batten
15. timber shakes
16. roofing membrane
17. SIP panel
18. acoustic spray
19. rafter as per S.E. spec
20. actuator, nominal 100x66mm
21. blind (N10/240) (Tess100 or 101)
22. insect mesh
23. electrical conduit see M&E spec
24. SW plate as per SE spec
25. steel ring beams as per SE spec
26. steel flat as per SE spec
27. lighting as per M&E spec
28. 18mm ply backing
29. fixing cleat
30. electrical containment zone
31. fixed window pane
32. glass
33. sliding window
34. fixed window beyond
35. window frame
36. timber ventilation screen
37. top hung ventilation panel
38. flitch plate as per SE spec
39. concrete floor edging tape and angle
40. concrete floor
41. underfloor heating
42. rigid insulation
43. DPM
44. concrete pontoon as per SE spec
45. shakes support batten
46. flat to support timber post as per SE spec
47. awning
48. 132x140 timber awning post
49. spine of external wall beyond
50. awning arms & bracket
51. window box
52. awning post
53. drip flashing fixed to window
54. DPC lapped to window
55. stainless steel bracket as per SE spec

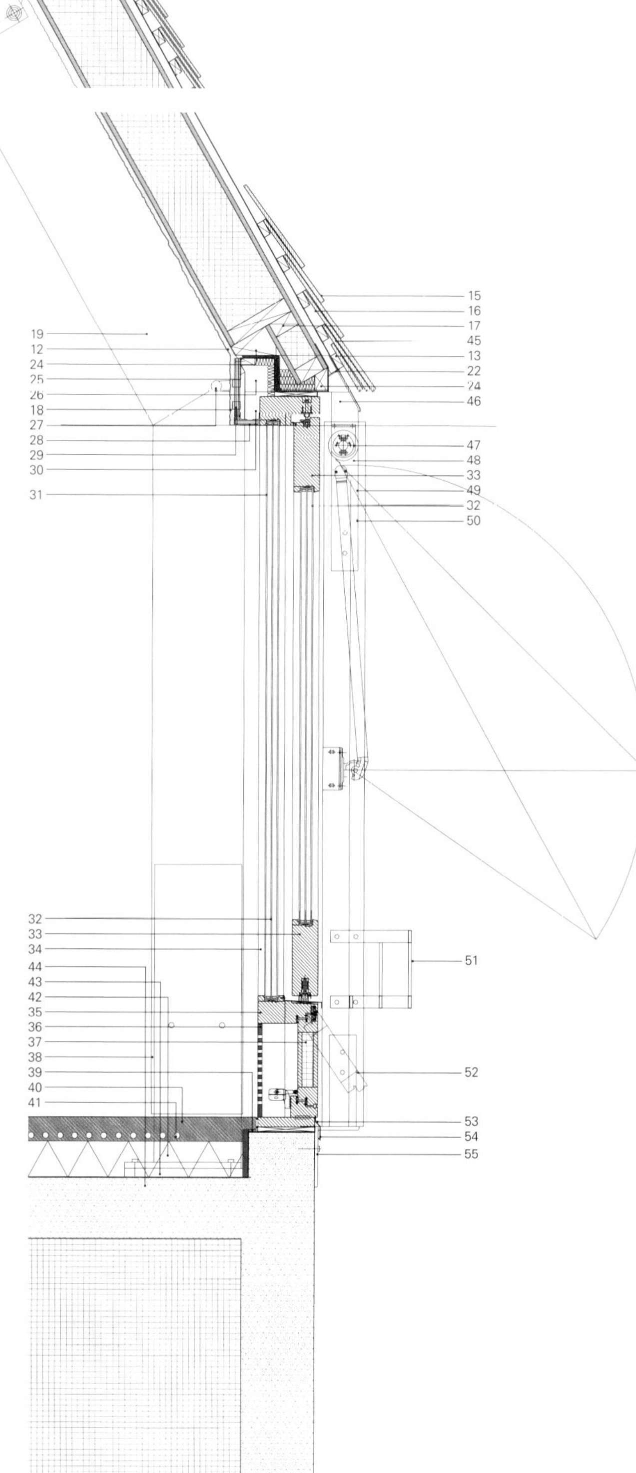

详图3 detail 3

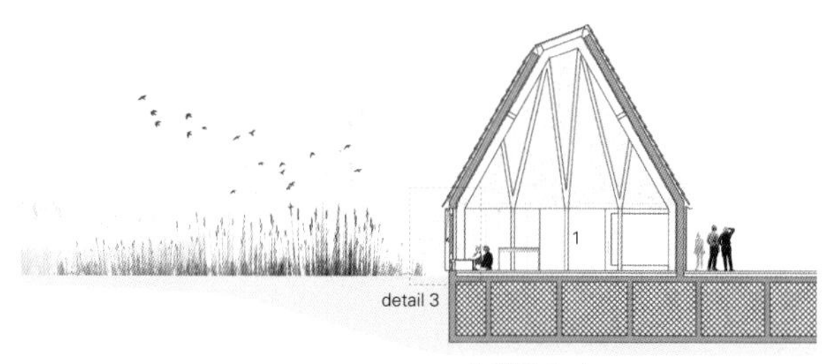

1 教室 1. classroom
B-B' 剖面图 section B-B'

项目名称：Brockholes Visitor Center
地点：Brockholes Nature Reserve, Preston, PR5 0UG, Lancashire, UK
建筑师：Adam Khan Architects
项目经理：Bovis Lend Lease，WDA Projects
结构工程师：Price & Myers
机械和电气工程师：Max Fordham LLP
承包商：Mansell PLC
工料测量师：Jackson Coles LLP
音响工程师：David Bonnett Associates
CDMC：Bovis Lend Lease
餐饮顾问：Tribourne
环境评估顾问：Scott Hughes
甲方：Lancashire Wildlife Trust
甲方项目经理：Ian Selby
浮桥面积：2,795m^2 / 建筑面积：1,400m^2 (internal)
承包价：GBP 6.25m
设计时间：2007 / 施工时间：2009.12—2011.6
摄影师：©Ioana Marinescu (courtesy of the architect)

Beautour中心的主要思想是使古老的乔治斯·杜兰德（旺代地区的一名博物学家，1886—1964）的府邸发扬光大，乔治斯·杜兰德拥有重要的收藏品，且迅速发展起一种对自然科学的激情。70年来，他在朋友、同行科学家的帮助下，从欧洲各地收集植物和昆虫。这就是他如何能够收集到近5000只鸟、150 000只蝴蝶和昆虫，以及众多标本的原因。因此，几乎他所有的4500种法国植物群都在此展出。

该项目的目的是发展以生物多样性为主题的教育和科学支持，并且为整个区域开发管理策略，展望进化前景。除了设置为主题园林外，该项目还采用了一些明显的行动，如堆肥、利用雨水进行浇灌，旨在帮助生物多样性的新形式在这处被遗弃了近30年的地区内再生。一些地块已达到高峰状态，且全球性干预提出了两种选择：要么整体保护，要么最小干预，以便实现一个全新的自然变化。其他的一些场地则相反，由于频繁的割草和放牧，一直处于生物缺乏的状态。为了使一个新的生态系统能够确立在一个长期的基础上，这些场地可以采用更高水平的干预策略。

Beautour博物馆和生物多样性研究中心

Agence Guinée Et Potin Architectes

这座博物馆和生物多样性研究中心试图在轻微举动（保护场地已有的生物多样性）和巨大举动（对生物多样性产生积极的影响）之间找到一个适当的平衡。因此，该项目既不是一个主题公园，也不是一个观赏花园。这真的是一个拥有特定场地的项目，受当地生物的多样性、地形和其他适合Beautour的因素所启发。访问行程的设计遵循逻辑性的、科学性的目标，带领客人来到田野和山谷，那里的自然野生环境满足Beautour古老的和新建的花园和草地的要求。

在目前的景观绿化背景下，项目展现了较强的特征，用当代创新的方式重新解读了传统的技术，通过采用完全覆盖墙壁和建筑屋顶的茅草外表皮来实现。竞标的效果图显示了材料的自然老化，随着季节更替褪色成灰色色调和暗色。由于其紧凑的形状会与杜兰德先生的府邸争辉，因此这个中心的建设将变得富有肌理，它拥抱着杜兰德先生的府邸，围绕着它，在场地蔓延开来，没有打破自然秩序。实木板栗树的树干也混淆了类似项目的整体形象。建筑作为一个分支，平躺在地上，是一处"建成的景观"，也是一个完善自然透视法的"新地貌"。建筑从原地升起，能够允许原地保留生物多样性，最大限度地减少基础工程对生物多样性的影响。项目缓慢升起，以展现青蛙和苍鹭的栖息池塘。附属的技术设备房屋被漆成黑色，里面设有更衣室和燃木锅炉。一个教学用的温室紧挨着该中心，位于场地的入口处。

Beautour Museum and Biodiversity Research Center

Main idea of Beautour Center is to glorify the historical Georges Durand's mansion(a Vendean naturalist, 1886~1964) who got important collections. He quickly developed a passion for natural sciences. For 70 years, he collected plants and insects from all over Europe, with the help of his friends and fellow scientists. This is how he has been able to collect nearly 5,000 birds, 150,000 butterflies and insects, and numerous herbariums. Thus almost all 4,500 species of the French flora are hereby represented.

The project aims to develop educational and scientific supports themed on biodiversity, as well as a management strategy and evolution prospectives for the whole area. Beyond the thematic gardens, composting, and using rainwater for watering that are some obvious actions, the project aims to help new forms of biodiversity to regenerate this site, abandoned for 30 years. Some plots of land have reached a state of climax, and the global intervention presents two alternatives: either an integral preservation,

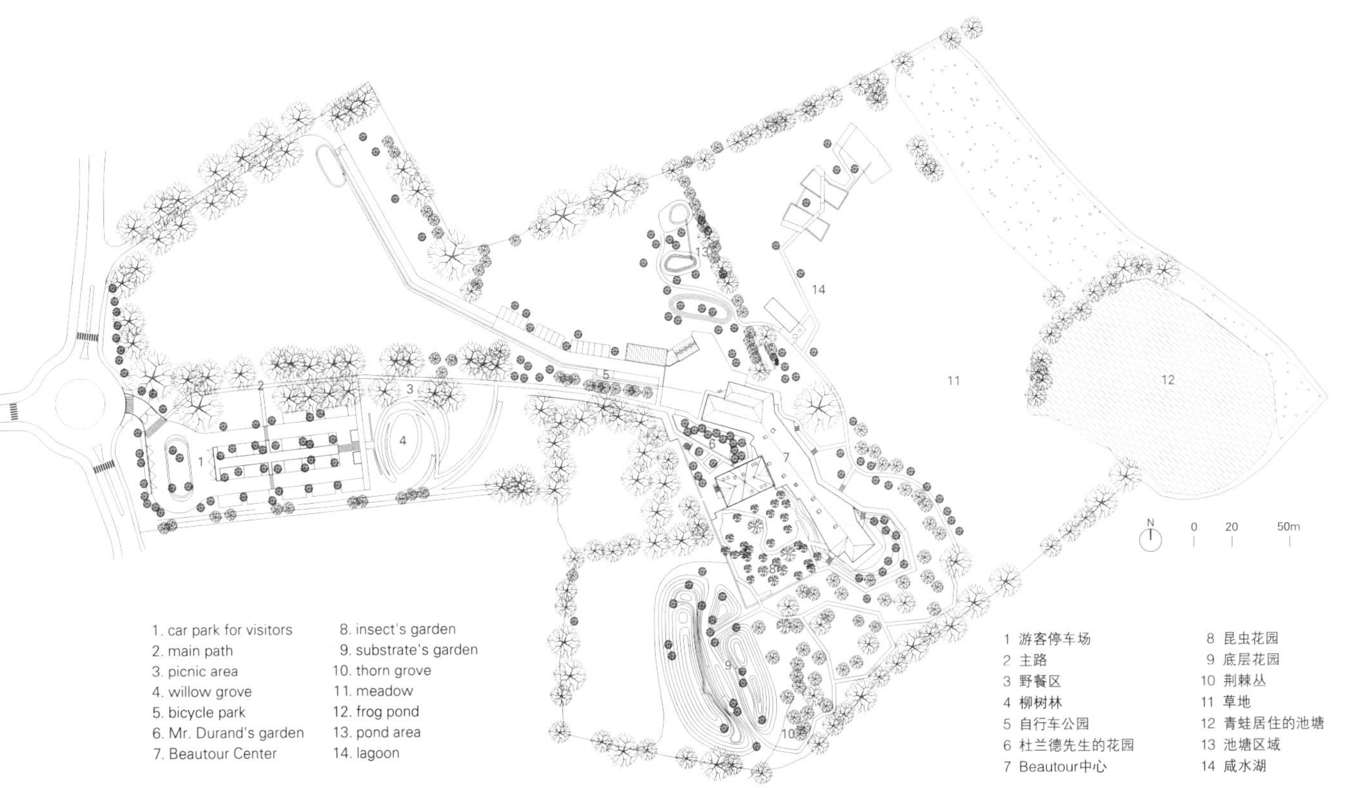

1. car park for visitors
2. main path
3. picnic area
4. willow grove
5. bicycle park
6. Mr. Durand's garden
7. Beautour Center
8. insect's garden
9. substrate's garden
10. thorn grove
11. meadow
12. frog pond
13. pond area
14. lagoon

1 游客停车场
2 主路
3 野餐区
4 柳树林
5 自行车公园
6 杜兰德先生的花园
7 Beautour中心
8 昆虫花园
9 底层花园
10 荆棘丛
11 草地
12 青蛙居住的池塘
13 池塘区域
14 咸水湖

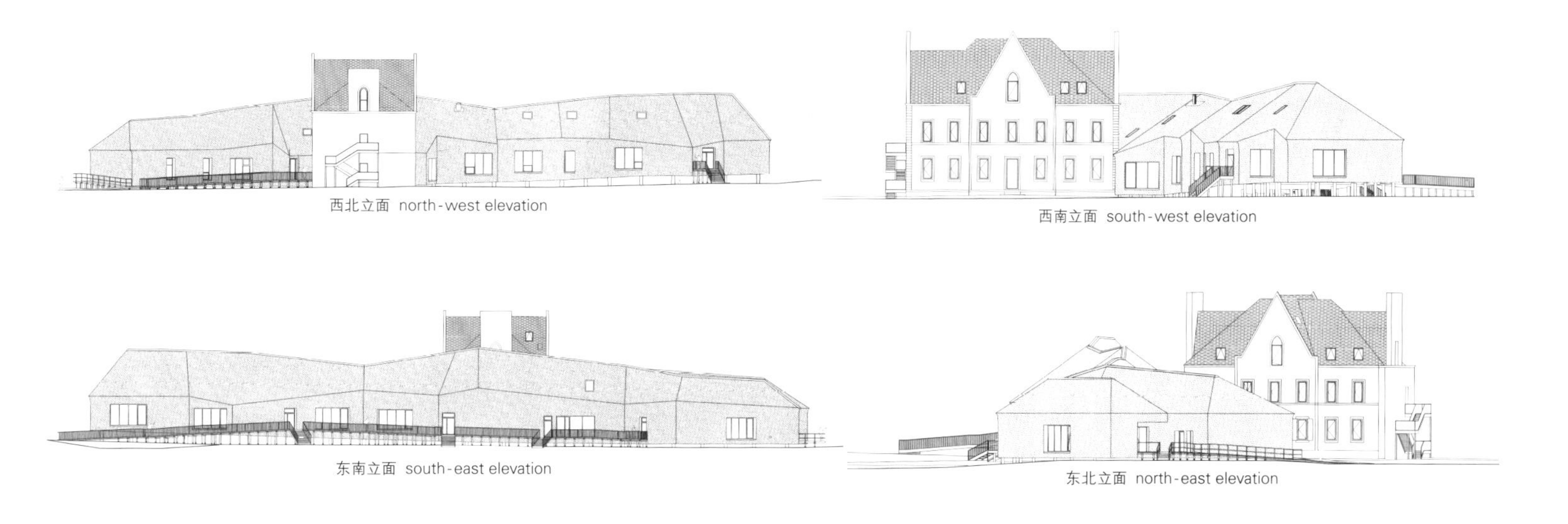
西北立面 north-west elevation

西南立面 south-west elevation

东南立面 south-east elevation

东北立面 north-east elevation

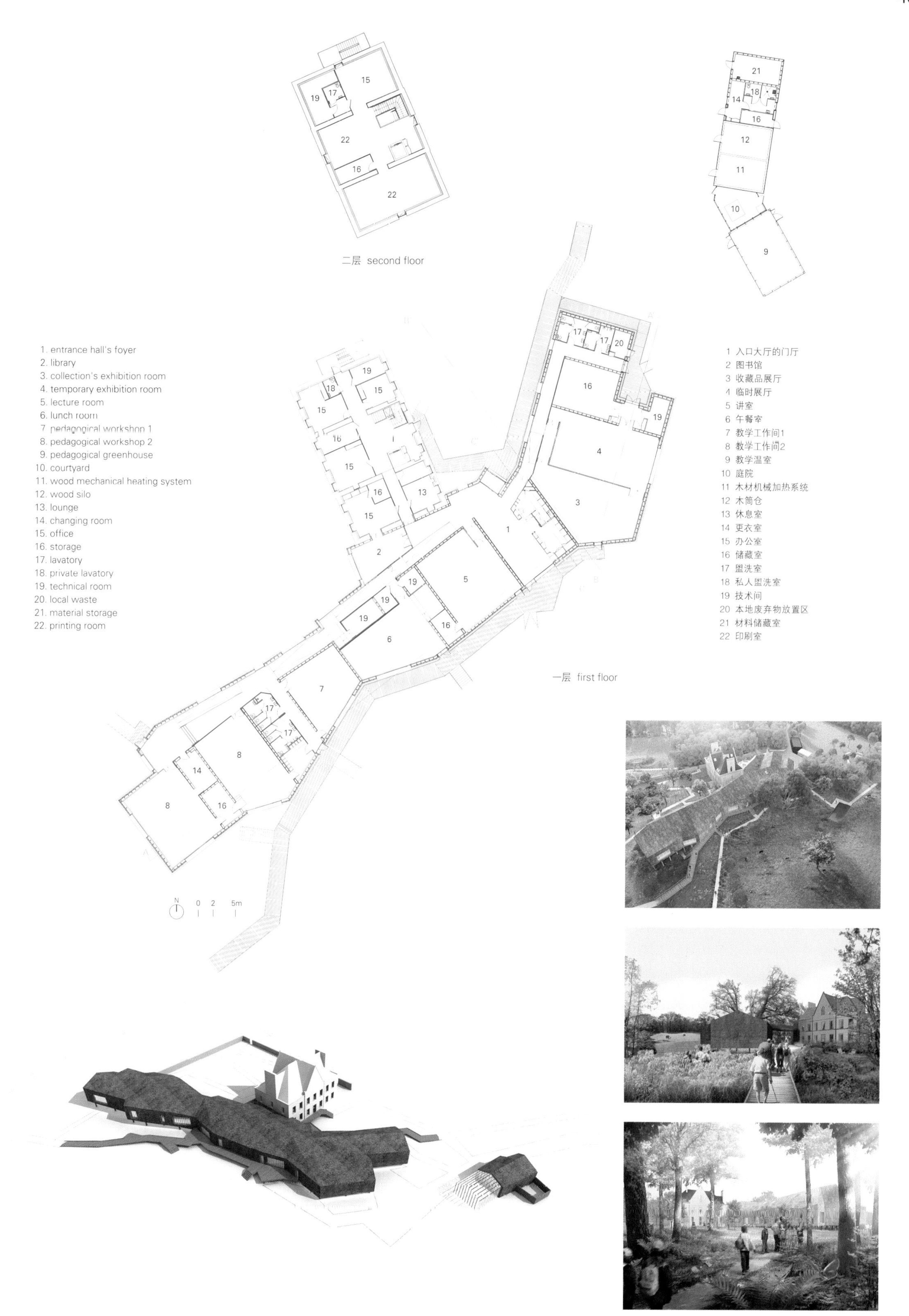

二层 second floor

一层 first floor

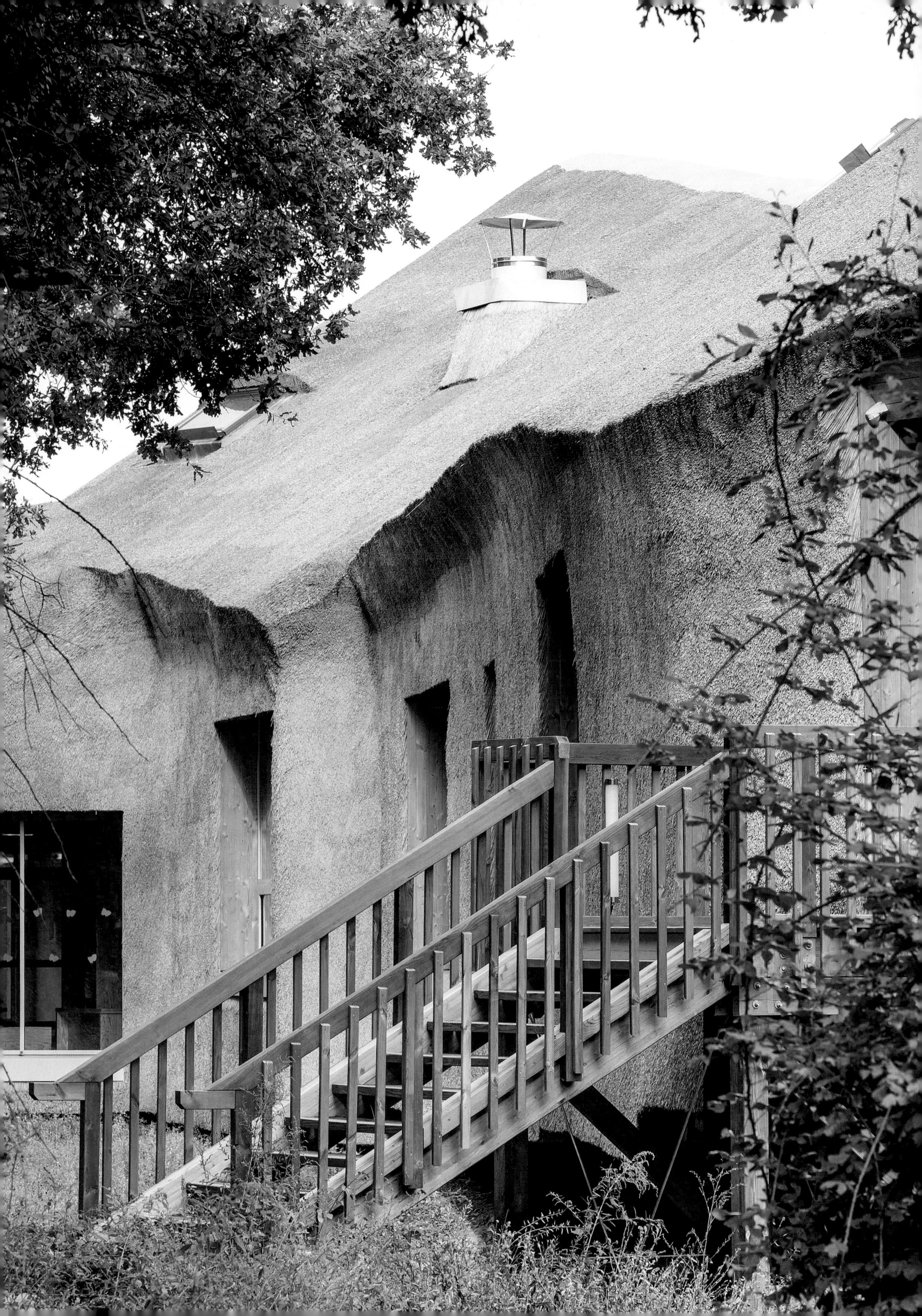

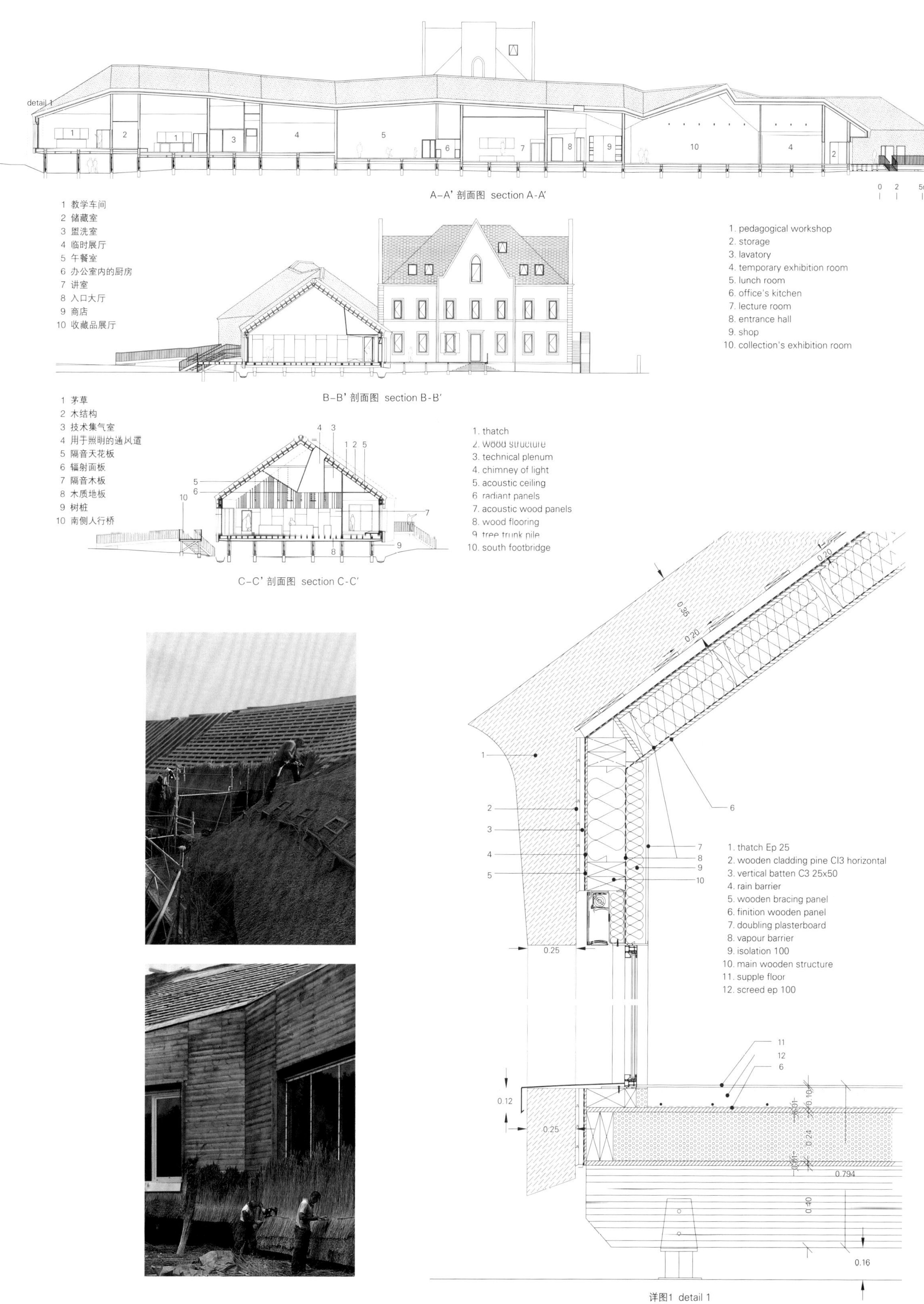
detail 1
1 教学车间
2 储藏室
3 盥洗室
4 临时展厅
5 午餐室
6 办公室内的厨房
7 讲室
8 入口大厅
9 商店
10 收藏品展厅
1. pedagogical workshop
2. storage
3. lavatory
4. temporary exhibition room
5. lunch room
6. office's kitchen
7. lecture room
8. entrance hall
9. shop
10. collection's exhibition room
0 2 5m
1 茅草
2 木结构
3 技术集气室
4 用于照明的通风道
5 隔音天花板
6 辐射面板
7 隔音木板
8 木质地板
9 树桩
10 南侧人行桥
1. thatch
2. wood structure
3. technical plenum
4. chimney of light
5. acoustic ceiling
6. radiant panels
7. acoustic wood panels
8. wood flooring
9. tree trunk pile
10. south footbridge
1. thatch Ep 25
2. wooden cladding pine Cl3 horizontal
3. vertical batten C3 25x50
4. rain barrier
5. wooden bracing panel
6. finition wooden panel
7. doubling plasterboard
8. vapour barrier
9. isolation 100
10. main wooden structure
11. supple floor
12. screed ep 100
0.35
0.20
0.25
0.12
0.794
0.40
0.16

A–A' 剖面图 section A-A'

B–B' 剖面图 section B-B'

C–C' 剖面图 section C-C'

详图1 detail 1

or a minimal intervention that could engage a new natural diversification. Some other plots, on the contrary, have been maintained in a state of biological poverty due to frequent mowing and pasture. These ones could use a higher level of interventionism, in order for a new ecosystem to settle on a long term basis.

This Museum and Biodiversity Research Center tries to find a right balance between light actions, preserving the biodiversity already on site, and other stronger actions, creating a positive impact on the biological diversity. Thus the project is neither a theme park, nor an ornamental garden. This really is a site-specific project, inspired by the local biodiversity, the topography, and the other qualities that are proper to Beautour. The visit itinerary is drawn by this logic, scientific purpose leading the visitor down to the fields and the valley, where the wild nature meets both Beautour historical and newly designed gardens and meadows.

In a very present landscaped green setting, the project takes on a strong identity, reinterpretating a traditional technique in a contemporary and innovative way, by adopting a thatched skin, that entirely covers both walls and roof of the building. The competition renderings display the natural aging of the material, fading to gray tones and shades as the seasons pass by. As a compact shape would have vied with Mr Durand's mansion, the building grows organic, embracing the mansion, surrounding it and spreading on the site without overthrowing the natural order. Solid raw chestnut tree trunks also confuse the overall image of the mimetic project. The building, as a branch laying on the ground, is a "piece of built landscape", a "new geography" completing the natural scenography. Making the building rise up from the ground allows the biodiversity to stay in place and minimizes the impact of foundation works. The project slowly lifts up to unveil the pond hosting frogs and herons. The technical facilities annex is painted black and houses lockerrooms and a wood-fired boiler. A pedagogical greenhouse stands next to it at the entrance of the site.

NATURALISME

VALLEES

项目名称：Museum and Biodiversity Research Center
地点：Le Bourg-sous-la-Roche, Beautour, La Roche sur Yon
建筑师：Agence Guinée Et Potin Architectes
设计团队：Anne-Flore Guinée, Solen Nico
透视图设计：Block Architectes
平面设计：WARMGREY
博物馆平面设计：Stéphanie VINCENT
结构和流体工程师：ISATEG
音效工程师：ITAC
景观建筑师：Guillaume Sevin Paysages
甲方：Région des Pays de la Loire
用地面积：05,000m^2
总建筑面积：2,057m^2
有效楼层面积：1,597m^2
造价：EUR 5,000,000
设计时间：2010
施工时间：2012—2013
摄影师：
©Stéphane Chalmeau (courtesy of the architect) - p.42, p.43, p.44, p.49, p.50[bottom], p.51
©Sergio Grazia (courtesy of the architect) - p.40~41, p.46, p.48
©Nicolas Pineau (courtesy of the architect) - p.50[top]

福古岛自然公园总部

Oto Arquitectos

在福古岛1800m的高空，火山的火山口，有一个约1200人的村庄，人们生活在法规的边缘处，占用国家的土地，主要从事农业活动，确保他们在佛得角的一处最贫困的地区生存下来。

国家利益保护区的现状迫使农业进行分区，限制施工，并且引进了反对对城镇进行自由占用的规则，因此产生利益的碰撞和频繁的冲突。这个项目来自于巩固一处受保护的区域的身份和运用新的公园管理措施来调节人口的需求。因此，建筑师创造了供居民和游客进行文化和娱乐享受的空间；这些工作区域也是通过雇佣技术人员来创造的，他们负责保护区的管理和维护工作。

将火山和火山口作为标志的自然景观是一种独特且罕见的美景，有潜力成为世界文化遗产。在这样的背景下，其基本思想是设计一座建筑，使之成为景观的一部分，且景观作为建筑物的一部分，并且其某些深色表皮进行了融合。在白天，长长的墙体塑造了建筑的外形，并且与道路相融合，创造出交织的混合阴影。在晚上，建筑没有使用强光，以保护本土的鸟类，所有照明都是间接的。所面临的当地资源短缺的挑战成为了机会，因此，该建筑使用当地的材料和技术，由人创造，且为人创造。

为了解决任何公用电网建设的不足，该建筑有太阳能做为主要来源，保证能源独立，以及双水电网；水来自于屋顶雨水和生活用水，由大型地下矿床得以补充，雨季过后，进行储存，一年一次。整座建筑被种有自然公园的代表性植物物种的、延伸至建筑外面并与周围环境相融合的斜坡和空间所包围。建筑分为两个区域：文化区——由一个带顶的礼堂、一个开放的礼堂、图书馆和露台酒吧组成；行政区则包括会议室、办公室、实验室和技术区。

由于自然公园的总部全面开始运作，因此公园的价值也开始上升，这有助于丰富小岛上的社会、文化和经济部分，并且开始以和谐的方式融入周围的空间，并对其进行突出。

WATER COLLECTION / WASTE MANAGEMENT

Rain water is received and directed along the top of the building to a storage tank, from which it can be used both as irrigation and domestic water. Grey waters are recollected, recycled and pumped back into system.

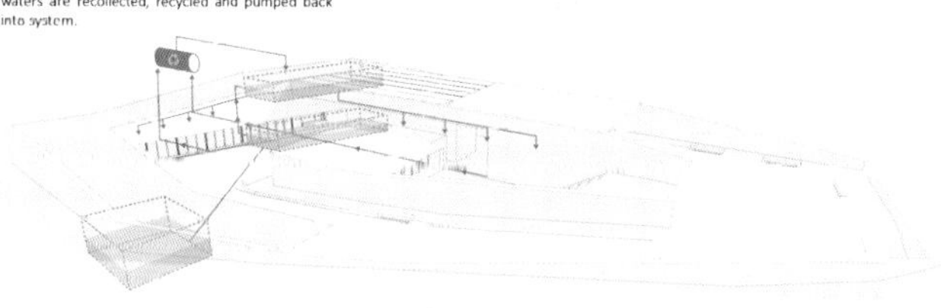

水 water

ENERGY PRODUCTION

Photovoltaic roof panels absorb solar energy which is then stored and/or delivered as electricity, covering the needs of the LED-based lighting system.

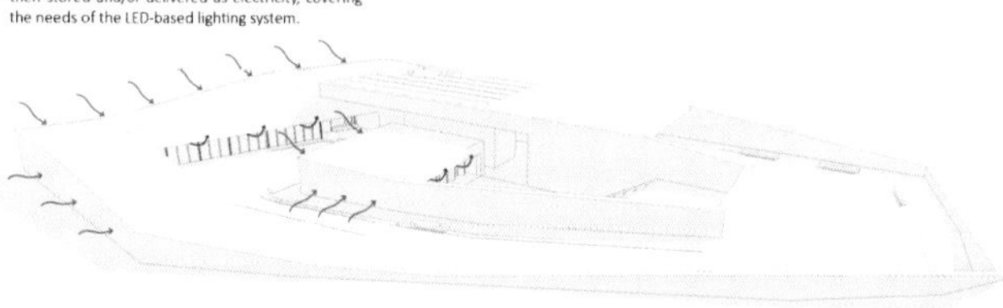

能源 energy

NATURAL VENTILATION

Facade integrated grid systems allow for a passive control of internal temperatures, taking advantage of the building's thermal inertia, which allows for heat accumulation during daytime and natural ventilation during the night.

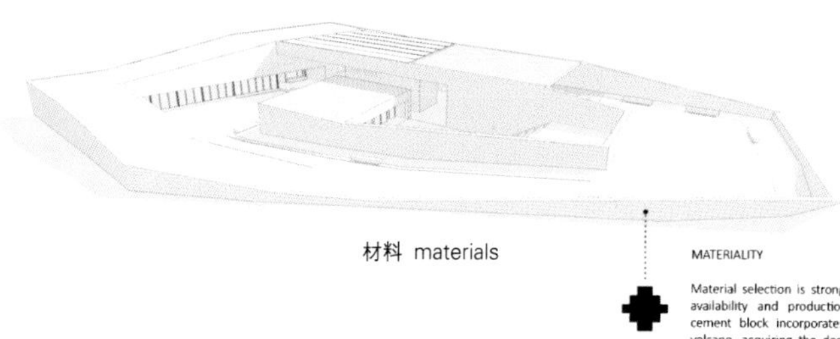

材料 materials

MATERIALITY

Material selection is strongly based on local availability and production: local masonry cement block incorporates ashes from the volcano, acquiring the desired dark tonality, blending building and landscape together.

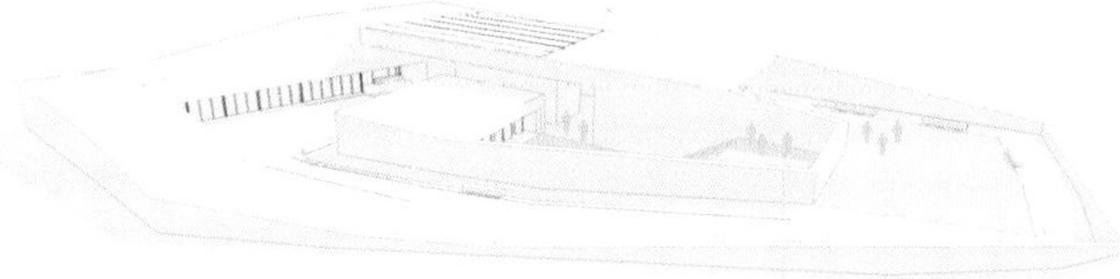

社交区域 social

SOCIAL INTERACTION

The local community is considered as a driving force for the project, which seeks to develop adequate spaces for social interaction. A caffè and internal/external auditoriums act as stages for cultural and recreational activities.

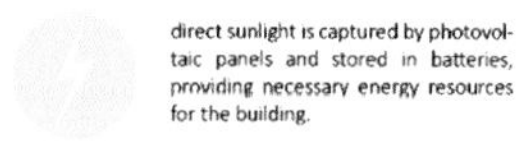

direct sunlight is captured by photovoltaic panels and stored in batteries, providing necessary energy resources for the building.

rain water is received and directed along the top of the building to a storage tank, from which it can be used both as irrigation and domestic water.

the local community is considered as a driving force for the project, which seeks to develop adequate spaces for social interaction, such as a bar and external auditorium.

material selection is strongly based on local availability and production: local production, for local people, aiming for local development.

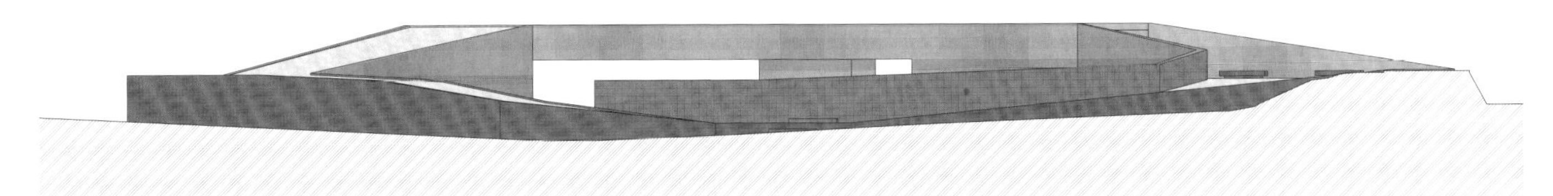

南立面 south elevation

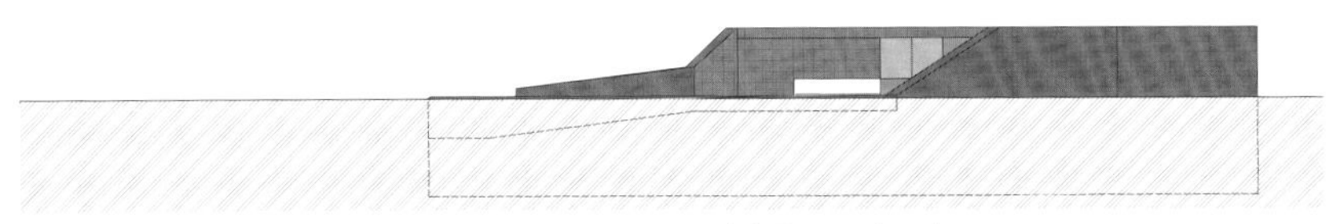

东立面 east elevation

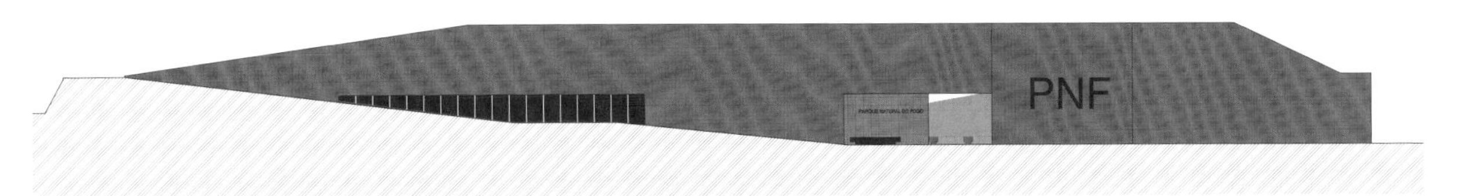

北立面 north elevation

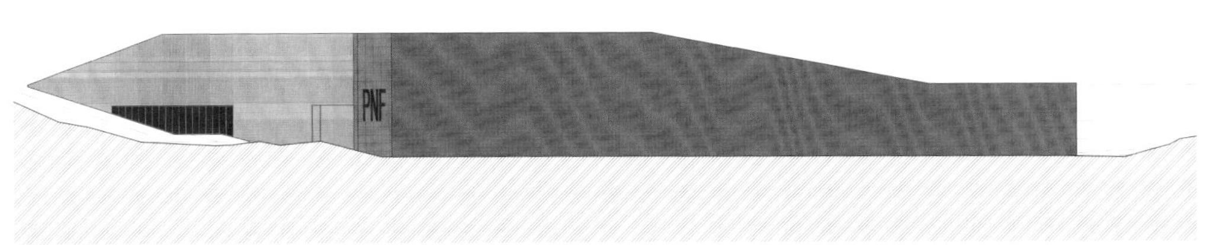

0 5 10m

西立面 west elevation

Fogo Island Natural Park Headquarters

On Fogo Island, at 1,800 meters of altitude, in the crater of the volcano, there is a village with about 1,200 people living on the fringes of legality, occupying lands of the state where they organize mainly agricultural activities, ensuring their survival in one of the poorest areas of Cape Verde.

The status of protected area of national interest forced the zoning of farming, with limitation to construction, and introduced rules against the free occupation of the town, generating collisions of interests with frequent clashes. This project was born from the need to consolidate the identity of a protected area and to conciliate the population with the new park management. Accordingly, the architect created spaces for cultural and recreational enjoyment for both residents and visitors; Workspaces were also created for employing technicians that will be responsible for the management and treatment of the protected area.

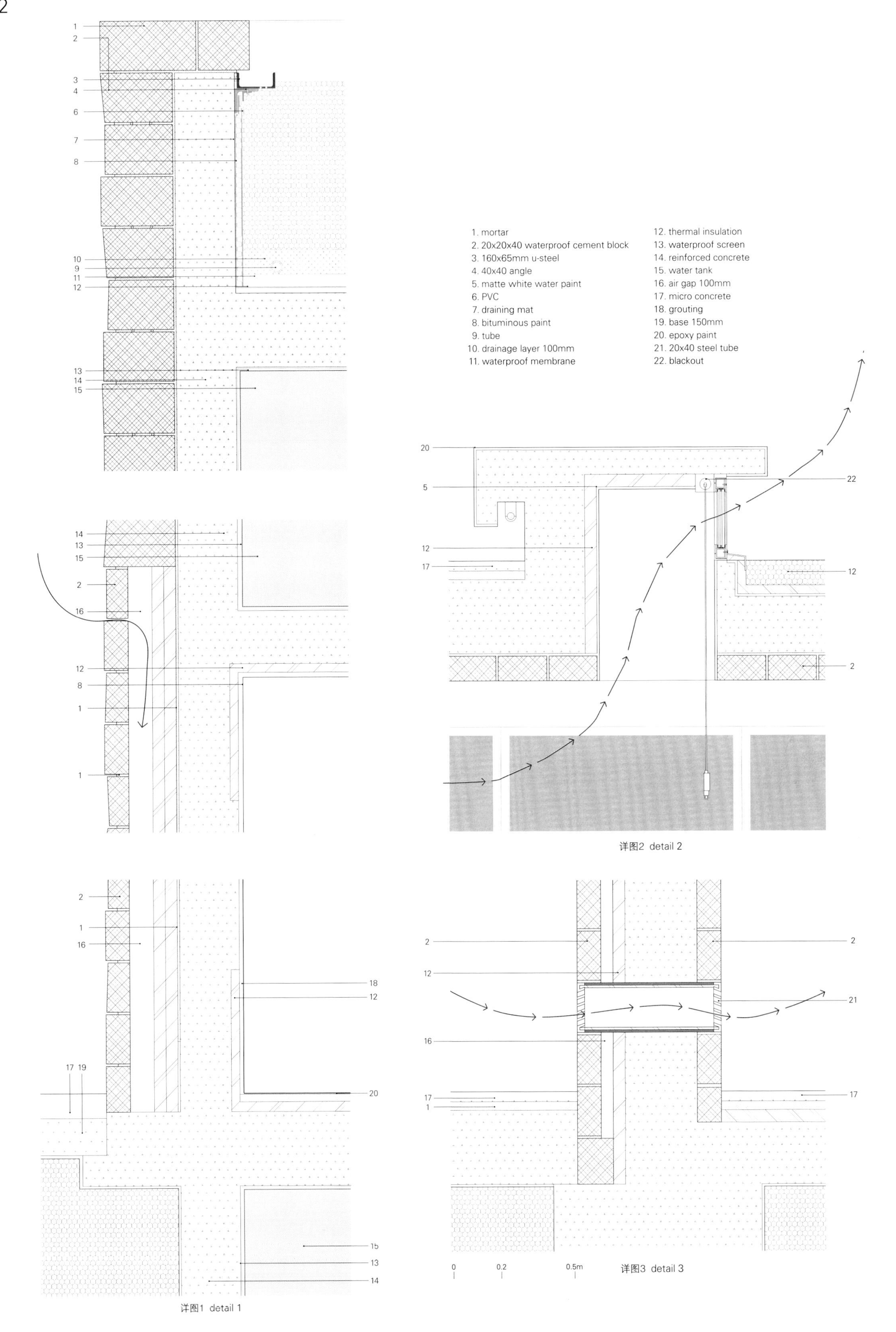

详图2 detail 2

详图3 detail 3

详图1 detail 1

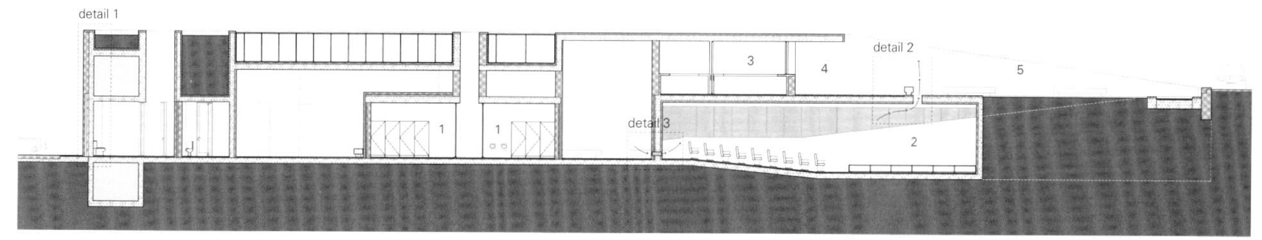

1 浴室 2 礼堂 3 酒吧 4 露台 5 风土花园
1. bathroom 2. auditorium 3. bar 4. terrace 5. endemic garden
A-A' 剖面图 section A-A'

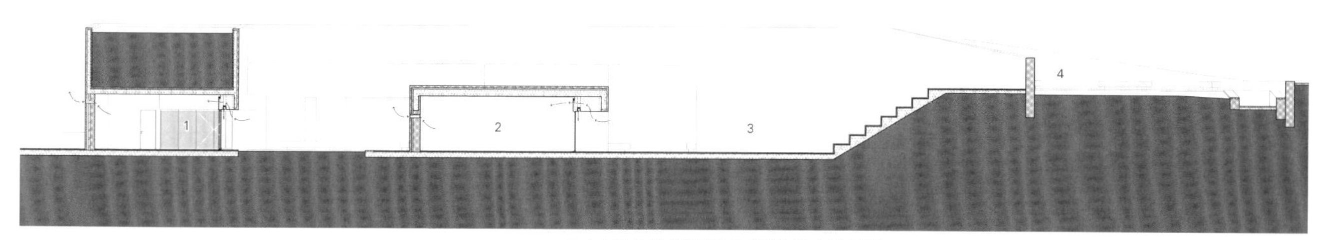

1 主入口 2 图书馆/展区 3 室外礼堂 4 风土花园
1. main entrance 2. library/exhibition 3. outdoor auditorium 4. endemic garden
B-B' 剖面图 section B-B'

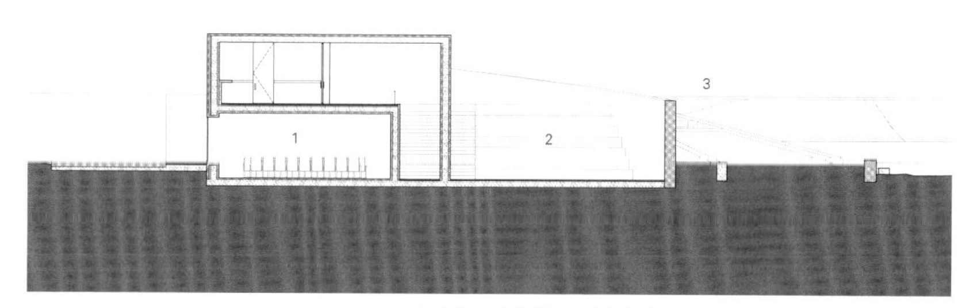

1 礼堂 2 室外礼堂 3 风土花园
1. auditorium 2. outdoor auditorium 3. endemic garden
C-C' 剖面图 section C-C'

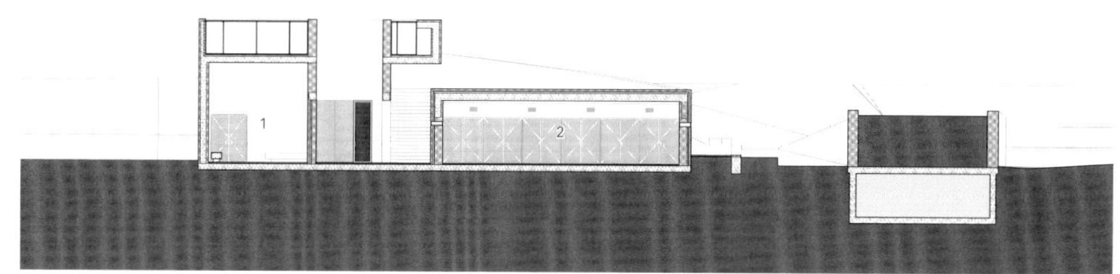

1 浴室 2 图书馆/展区
1. bathroom 2. library/exhibition
D-D' 剖面图 section D-D'

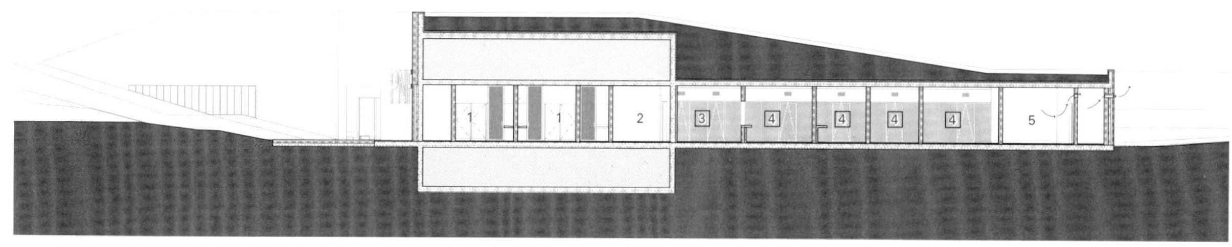

1 浴室 2 主入口 3 接待处 4 实验室 5 会议室
1. bathroom 2. main entrance 3. reception 4. laboratory 5. meeting room
E-E' 剖面图 section E-E'

0 5 10m

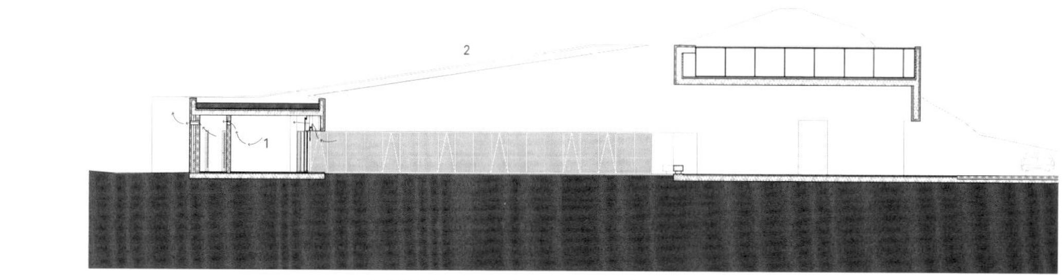

1 主管办公室 2 风土花园 1. director office 2. endemic garden
F-F' 剖面图 section F-F'

项目名称：Fogo Island Natural Park Headquarters
地点：Ch das Caldeiras, Ilha do Fogo - Cabo Verde
建筑师：Oto Arquitectos
项目团队：André Castro Santos, Miguel Ribeiro de Carvalho,
Nuno Teixeira Martins, Ricardo Barbosa Vicente
专家：Amado Alves, Maria João Rodrigues + João Parente, Prosirtec, Matriz Engenharia
承包商：Armando Cunha
甲方：Ministry of Agriculture
用地面积：3,200m² / 总建筑面积：701.21m² / 有效楼层面积：2,579.49m²
竣工时间：2013
摄影师：©FG+SG Architectural Photography

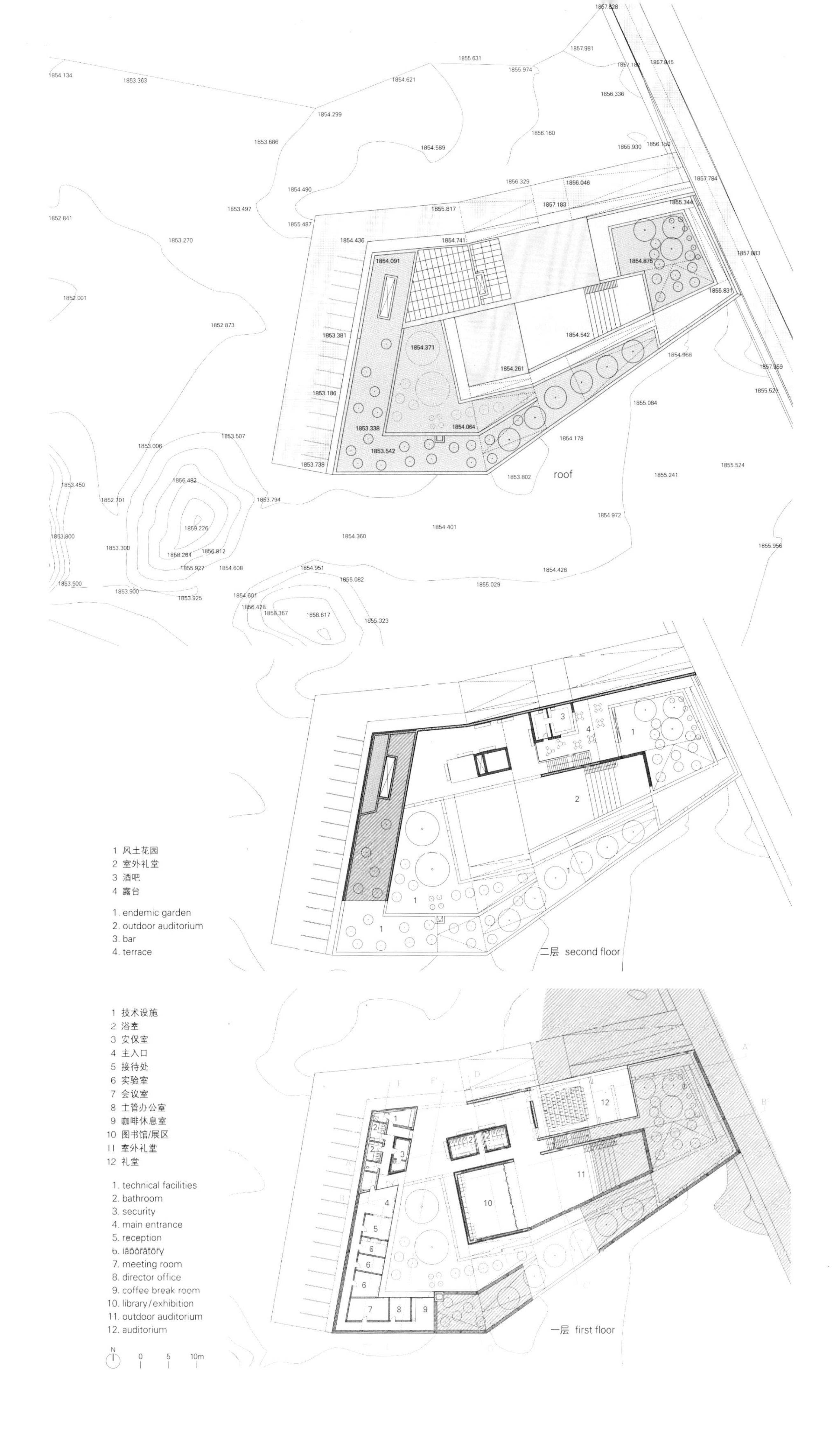
roof
1 风土花园
2 室外礼堂
3 酒吧
4 露台
1. endemic garden
2. outdoor auditorium
3. bar
4. terrace
二层 second floor
1 技术设施
2 浴室
3 安保室
4 主入口
5 接待处
6 实验室
7 会议室
8 主管办公室
9 咖啡休息室
10 图书馆/展区
11 室外礼堂
12 礼堂
1. technical facilities
2. bathroom
3. security
4. main entrance
5. reception
6. laboratory
7. meeting room
8. director office
9. coffee break room
10. library/exhibition
11. outdoor auditorium
12. auditorium
一层 first floor
N
0 5 10m

The natural landscape, deeply marked by the volcano and its crater, is a unique and rare beauty, with the potential to become a world heritage site. In this context, the basic idea was to design a building so as to be part of the landscape and the landscape being part of the building, and to have sort of a melting of darker skins. During daytime, the long walls shape the building and blend with the road creating a maze and a mix of shadows. At night, bright light is avoided, so to protect the native birds, and all lighting is indirect. The challenges of shortage of local resources became an opportunity and, therefore, the building was made by the people and for the people, using local materials and techniques.

To address the lack of any public utility grids, the building has a guaranteed energy independence with solar energy as the main source and a double water grid; water is taken from the roofs and daily used water is supplemented with large deposits, stocked annually after the rainy season. The whole building is surrounded by

ramps and spaces with representative plant species of the natural park extending to the exterior of the building and merging it with the surroundings. The building is divided into two zones: Cultural Zone – composed of a covered auditorium, an open auditorium, library and terrace bar; Administrative Zone – comprises meeting rooms, offices, laboratory and technical areas.

With the Headquarters fully operational, the Natural Park is increasingly valued which contributes to enrich the social, cultural and economic sectors of the island, starting to integrate and enhance in a harmonious way the surrounding space.

Sancaklar清真寺位于伊斯坦布尔郊区的Buyukçekmece街区，旨在通过基于形式和只关注宗教空间实质，来远离当下的建筑讨论方法，并解决设计一座清真寺的基本问题。

该项目选址于大草原景观，一条繁忙的高速公路将草原与周边郊区的门户社区隔离。围绕在清真寺上层庭院里的公园周围的高墙描绘了混乱的外部世界和公共公园的宁静气氛之间的明确界限。长长的天篷从公园伸出，成为从外部可见的唯一的建筑元素。建筑位于这个天篷之下，人们可以穿过花园从上层庭院的一条小径进入，该建筑完全融合于地形中，当人们穿过这些风景，下山且穿过石墙进入清真寺时，已经把外界遗留在世外。

清真寺的内部如同一个简单的洞穴，成为一处引人注目的、令人敬畏的空间，来进行祈祷，或与神灵独处。沿Qiblah方向的墙面的缝隙和裂缝突出了祈祷空间的方向性，并且允许光线渗入到祈祷大厅。

该项目一直展现了人类和自然之间的张力。遵循着景观的自然坡度的石质台阶和跨度超过6m的薄钢筋混凝土板（形成天篷）之间的对比有助于加强这种双重关系。

Sancaklar Mosque

Sancaklar Mosque located in Buyukçekmece, a suburban neighborhood in the outskirts of Istanbul, aims to address the fundamental issues of designing a mosque by distancing itself from the current architectural discussions based on form and focusing solely on the essence of religious space.

The project site is located in a prairie landscape that is separated from the surrounding suburban gated communities by a busy highway. The high walls surrounding the park on the upper courtyard of the mosque depict a clear boundary between the chaotic outer world and the serene atmosphere of the public park. The long canopy stretching out from the park becomes the only architectural element visible from the outside. The building is located below this canopy and can be accessed from a path from the up-

Sancaklar清真寺

Emre Arolat Architects

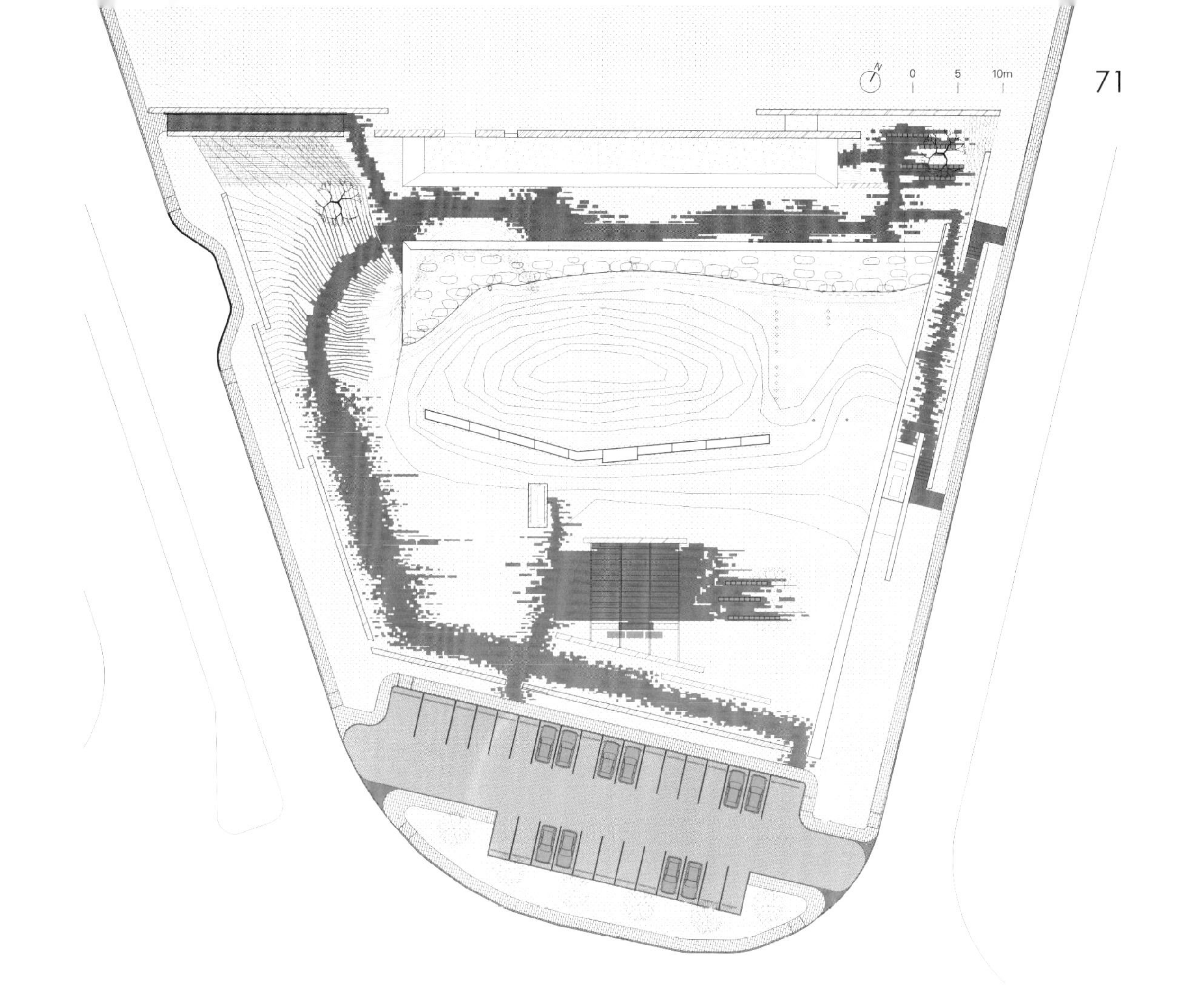
N
0
5
10m

07.12.2010
Sancak Camii

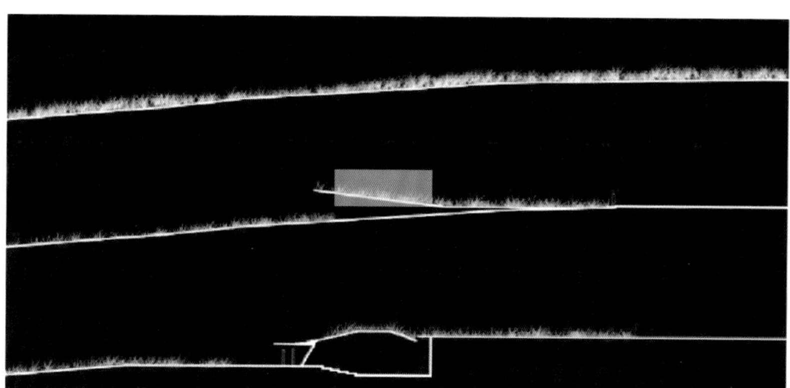

per courtyard through the park. The building blends in completely with the topography and the outside world is left behind as one moves through the landscape, down the hill and in between the walls to enter the mosque.

The interior of the mosque, a simple cave like space, becomes a dramatic and awe inspiring place to pray and be alone with God. The slits and fractures along the Qiblah wall enhance the directionality of the prayer space and allow daylight to filter into the prayer hall.

The project constantly plays off of the tension between man-made and natural. The contrast between the natural stone stairs following the natural slope of the landscape and the thin reinforced concrete slab spanning over 6 meters to form the canopy helps enhance this dual relationship.

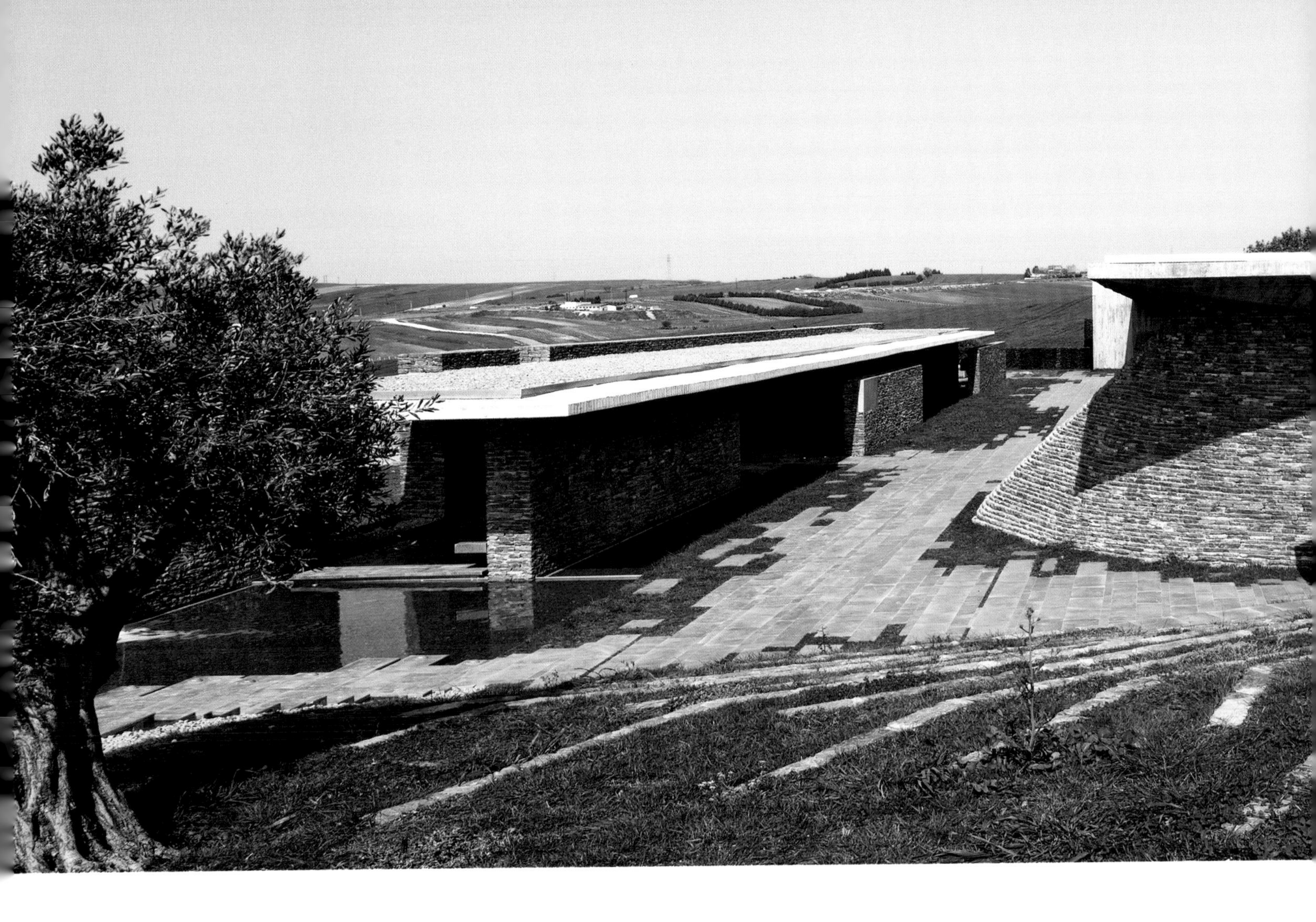

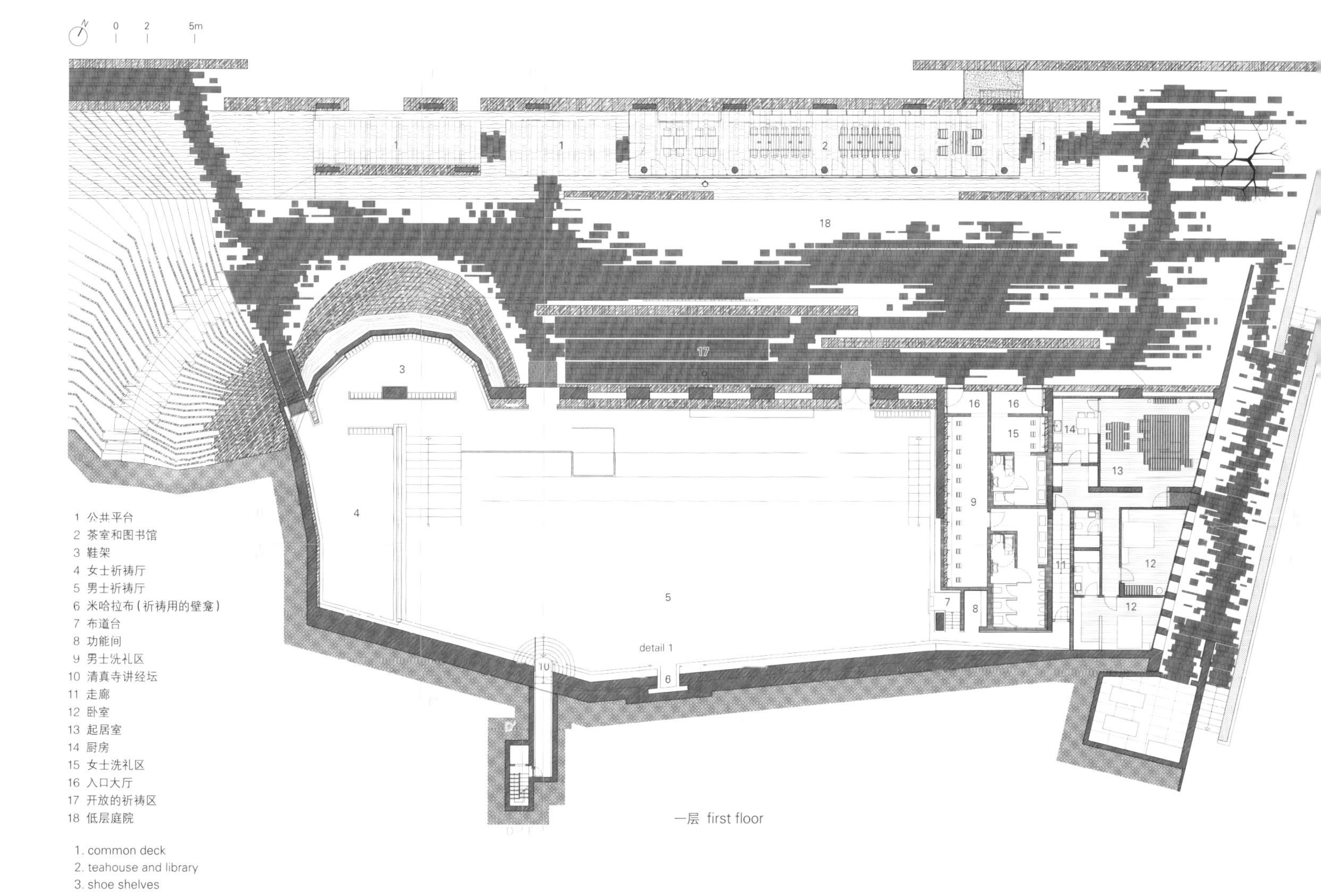

一层 first floor

1. common deck
2. teahouse and library
3. shoe shelves
4. women prayer's hall
5. men prayer's hall
6. mihrab(prayer niche)
7. sermonize podium
8. utility space
9. ablution area for men
10. minbar(pulpit)
11. corridor
12. bedroom
13. living room
14. kitchen
15. ablution area for women
16. entrance hall
17. open praying area
18. lower courtyard

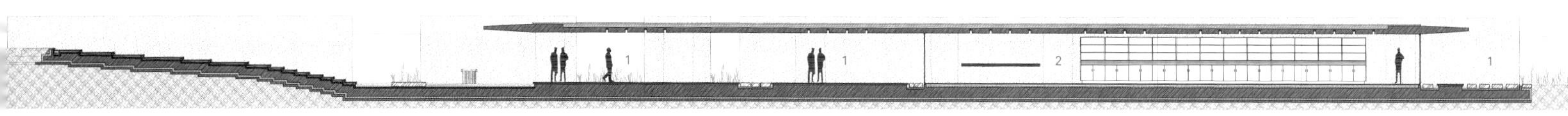

1 公共平台 2 茶室和图书馆
1.common deck 2. teahouse and library
A-A' 剖面图 section A-A'

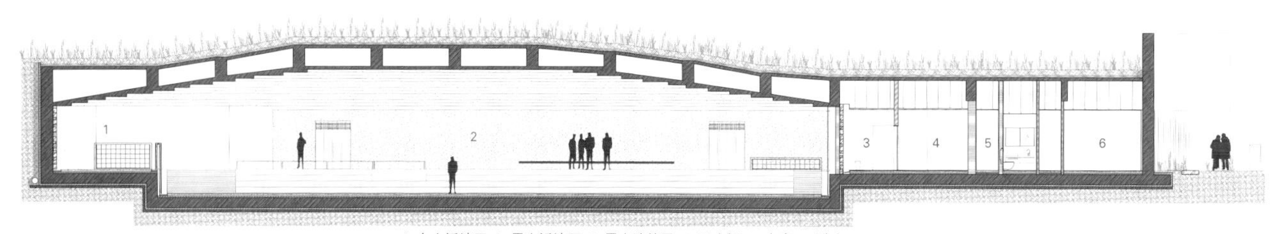

1 女士祈祷厅 2 男士祈祷厅 3 男士洗礼区 4 卫生间 5 走廊 6 卧室
1. women prayer's hall 2. men prayer's hall 3. ablution area for men 4. w.c. 5. corridor 6. bedroom
B-B' 剖面图 section B-B'

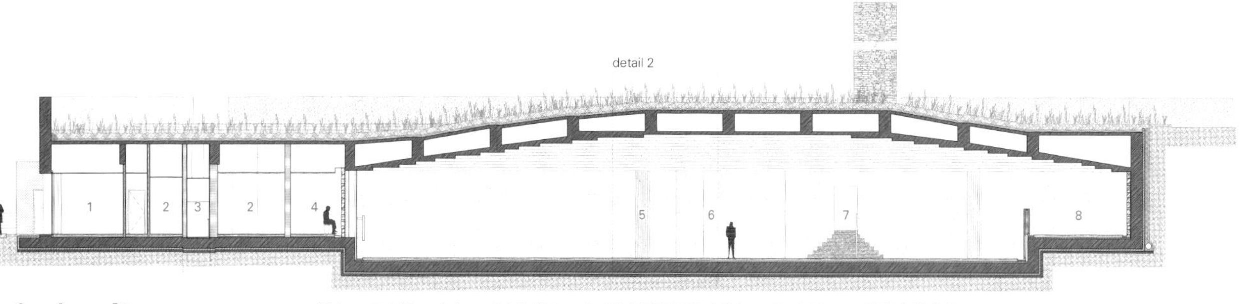

1 卧室 2 卫生间 3 走廊 4 男士洗礼区 5 米哈拉布(祈祷用的壁龛) 6 男士祈祷厅 7 清真寺讲经坛
1. bedroom 2. w.c. 3. corridor 4. ablution area for men 5. mihrab(prayer niche) 6. men prayer's hall 7. minbar(pulpit)
C-C' 剖面图 section C-C'

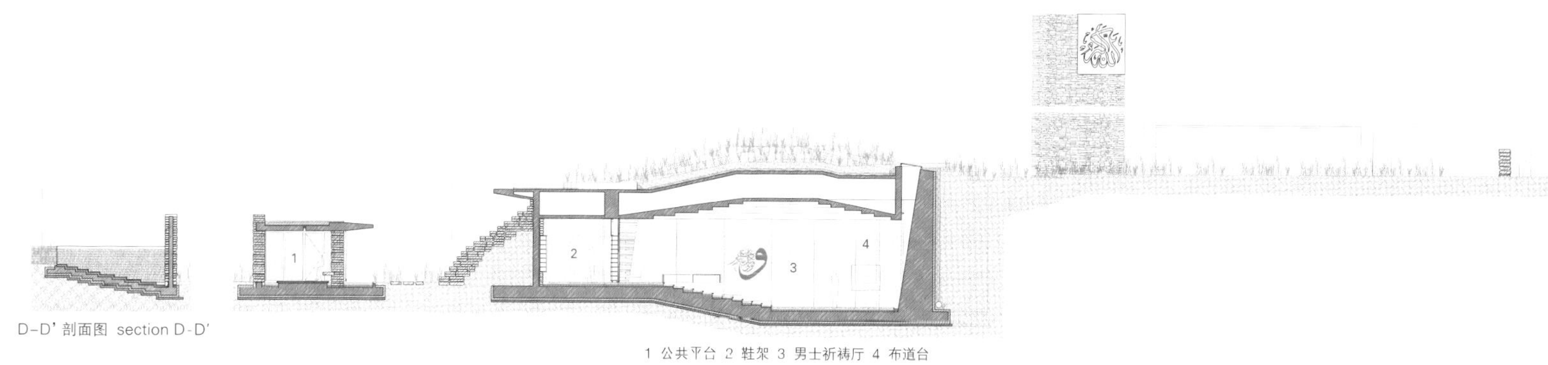

D-D' 剖面图 section D-D'

1 公共平台 2 鞋架 3 男士祈祷厅 4 布道台
1. common deck 2. shoe shelves 3. men prayer's hall 4. sermonize podium
E-E' 剖面图 section F-F'

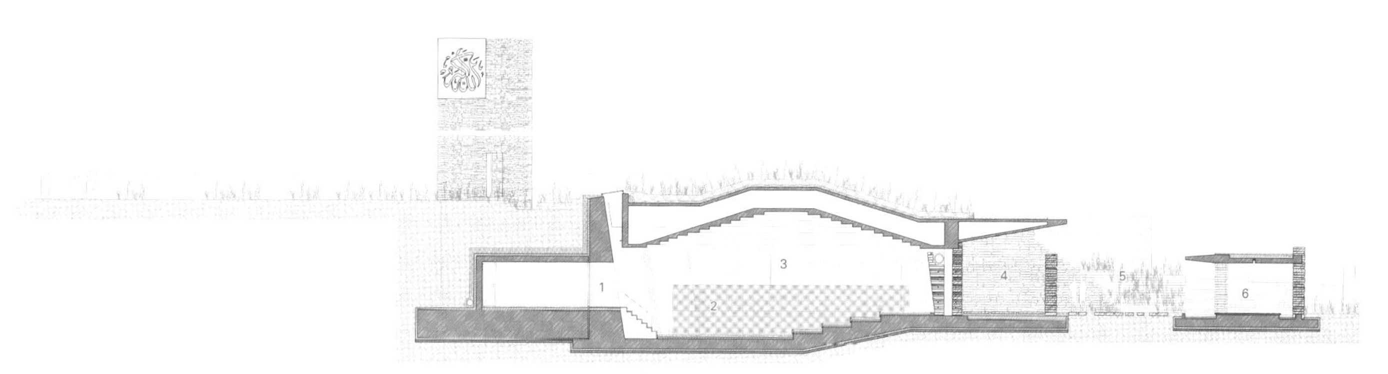

1 清真寺讲经坛 2 男士祈祷厅 3 女士祈祷厅 4 开放的祈祷区 5 低层庭院 6 公共平台
1. minbar(pulpit) 2. men prayer's hall 3. women prayer's hall 4. open praying area 5. lower courtyard 6. common deck
F-F' 剖面图 section F-F'

واذكر ربك كثيرا

项目名称：Sancaklar Mosque
地点：Istanbul, Turkey
建筑师：Emre Arolat Architects
项目团队：conceptual project_Emre Arolat, Leyla Kori, Nil Aynalı/ architectural project_Uygar Yüksel / Ayça Yontarım, Ayşegül Taşkın, Cem, Şahin, Fatih Tezman, Gönül Karahan, Nurdan Gürlesin, Oya Eskin Güvendi, Taha Alkan, Ünal Ali Özger
用地面积：700m²/总建筑面积：1,200m²/有效楼层面积：7,365m²
设计时间：2011/竣工时间：2012
摄影师：
©Cemal Emden(courtesy of the architect)-p.70~71, p.72, p.78top, p.79, p.80top, p.81top, p.82~83, p.84, p.86top
©Thomas Mayer(courtesy of the architect)-p.73, p.74~75, p.76~77, p.78bottom, p.80~81, p.85, p.86bottom

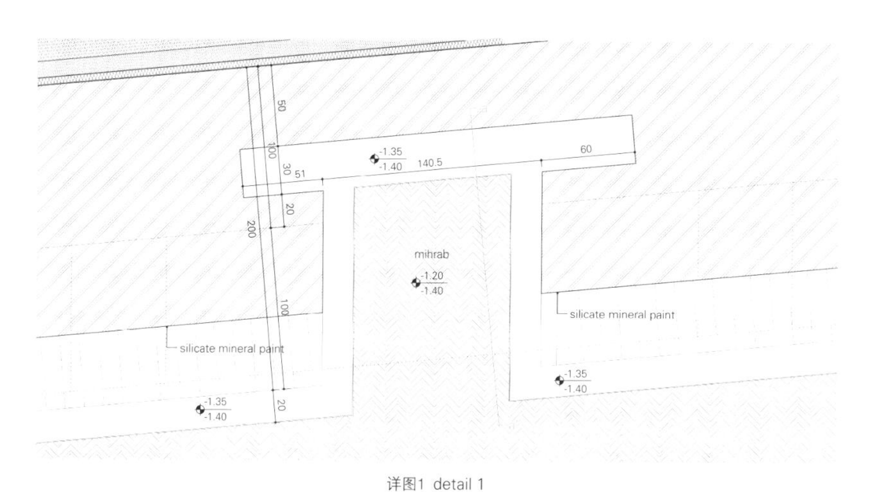

详图1 detail 1

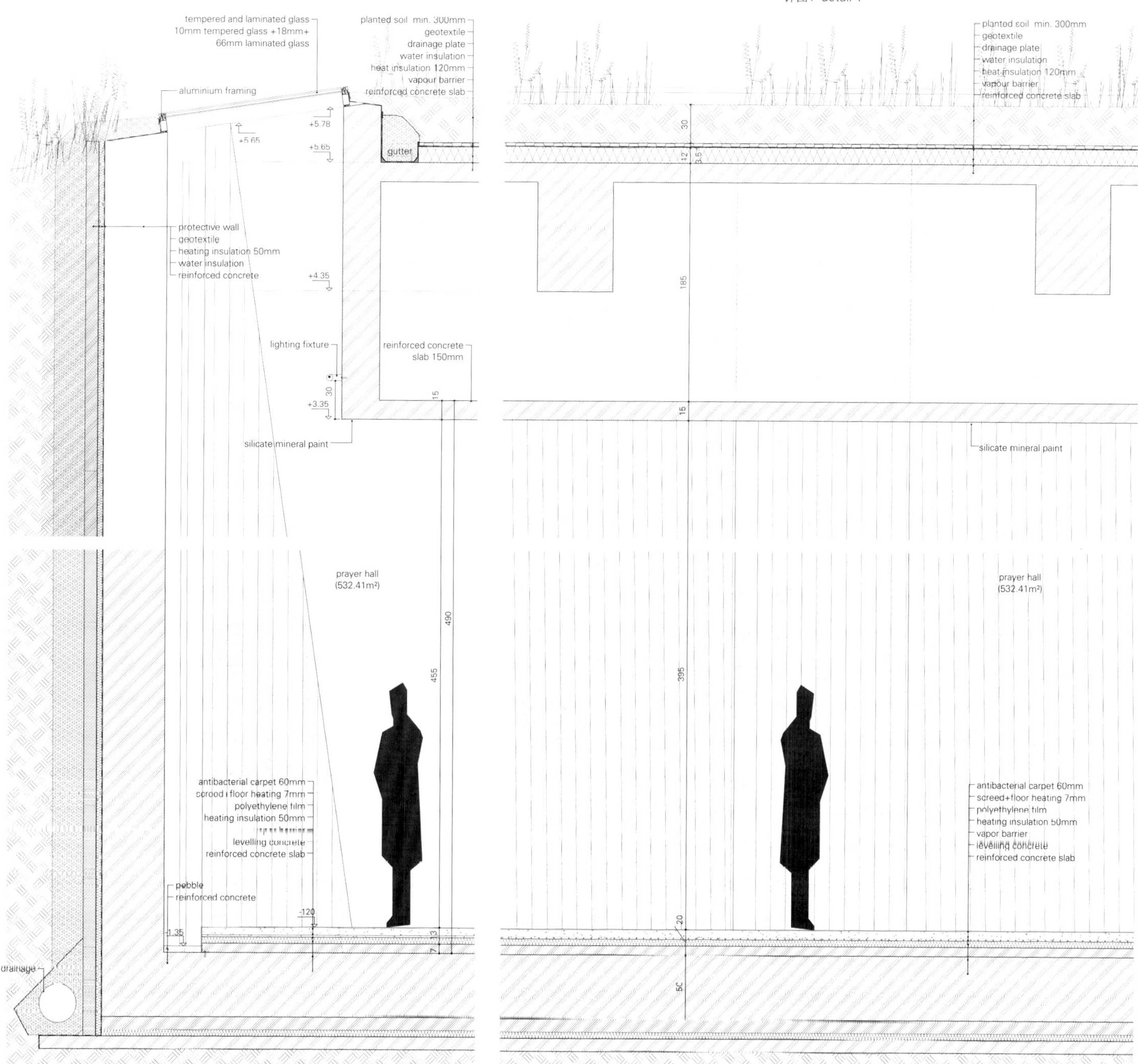

a-a' 剖面图 section a-a'

详图2 detail 2

玻璃农场

MVRDV

第二次世界大战期间，Schijndel市场广场遭到了市场花园行动的损害，此后一直经受数次的扩建和翻修。建筑师Winy Maas于1980年写了一封信，并于2000年在市议会上通过了一项在教堂、市政厅和主要街道之间的广场上建造新结构的想法。MVRDV事务所从那时起反复地提出了新的规划，来填补这个不寻常的大型乡村广场的空白。玻璃农场是MVRDV为这一场地提出的第七个建议。

该村庄积极地参与整个过程，并产生激烈的讨论、民意调查，以及由支持者和反对者在当地进行的媒体论战。完全覆盖玻璃幕墙的1600m^{2}的建筑物主要由一系列的公共设施组成，如餐厅、商店和一个健身中心。

巧合的是，由城市规划师所设计的最大的外围护结构有着传统的Schijndel农场的形式。所有遗留下来的本地农场都进行了测定、分析，并

且规划师从这些数据中设想一个"理想的"平均数。在与MVRDV的合作中，艺术家弗兰克·范·德·萨尔姆拍下所有的现存的传统农场，并从这些农场当中组成一个"典型农场"的图像。这个图像采用烧结方式印到1800m^2的玻璃立面上，效果就像大教堂的染色玻璃窗。根据光线和视野的需求，印制图像或多或少呈半透明状态。

到了晚上，结构会从内部照亮，成为农场的纪念碑。玻璃农场高14m，故意设计得不成比例，比一座真正的农场大1.6倍，象征着村庄成长为一个镇。印出的图像遵循这个"扩张的历史"，而叠加的农场大门为4m高。当成年人与建筑物产生互动时，他们可以再次体验儿时的感觉，可能会为他们接受这座建筑的情怀增加怀旧回忆的元素。为了进一步加强这一点，大楼旁会有一张桌子和一个秋千，以扩大农场庭院的规模。

Glass Farm

Schijndel's market square suffered from Operation Market Garden's damages during the Second World War and has been subject to numerous enlargements and refurbishments. Winy Maas wrote a letter in 1980, and in 2000 the town council adopted the idea of a new structure in the square between the church, town hall and main street. MVRDV since then iteratively proposed new options that could fill the gap of this unusually large village square. The Glass Farm is MVRDV's seventh proposal for the site.

The village engaged vividly in the process resulting in heated debates, polls and polemics in the local press – by supporters and

项目名称：Glass Farm
地点：Schijndel, Netherlands
首席建筑师：MVRDV
设计团队：Winy Maas, Jacob van Rijs, Nathalie de Vries, Frans de Witte, Gijs Rikken
结构：Hooijen Konstruktiebureau
安装：IOC Ridderkerk / 立面：Brakel Atmos / 玻璃：AGC
甲方：RemBrand bv
功能：retail, office, restaurant, wellness center / 面积：1,600m²
初始设计时间：2000 / 草图设计时间：2008— 2012 / 竣工时间：2013.1
摄影师：
courtesy of the architect - p.94[bottom], p.98
©Persbureau van Eijndhoven(courtesy of the architect) - p.90~91
©Thomas Mayer - p.88~89, p.95, p.96, p.97
©Daria Scagliola - p.92, p.94[top], p.99

农场式村庄（1832）
The farming village (1832)

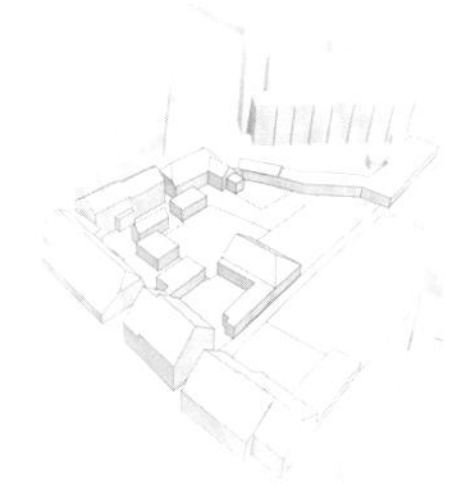

Before and during the Second World War, Schijndel did not have a market square. There were farms, houses and little shops, there were also yards, gardens and paths.

空地式广场（1954）
An empty square (1954)

The damage of the war and a certain drift for progress, led to a complete demolition of the centre, surrounded by new buildings. A large but empty square appeared.

毫无生气的场地（1954—2002）
Cheerless (1954~2002)

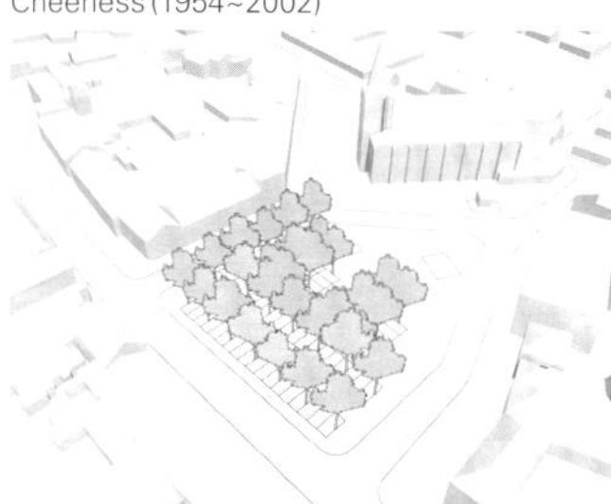

Has this been a good idea? Years of discussions and trying passed by... Trees or no trees? A fountain or not? Despite of all the efforts, the square didn't become any cozier. Suspiciously, they looked around to the surrounding villages. Should there be a pavilion on the market?

一座塔！（2002）
A tower! (2002)

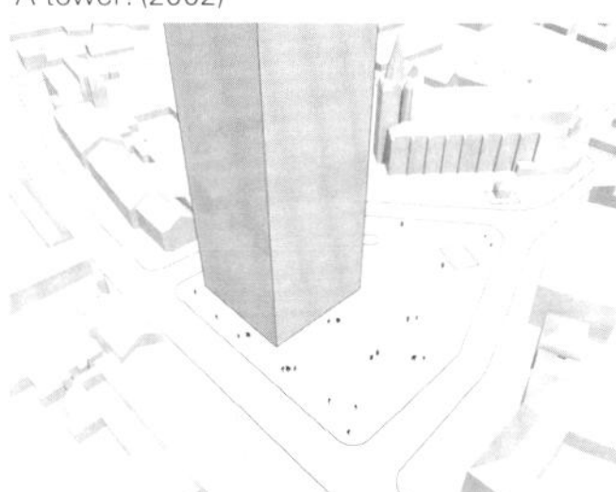

In 1980, 21 year old Winy Maas, born and raised in Schijndel, wrote a letter to the mayor. He asked if they could do something about the situation of the market square. The mayor answered he would come back to it. Decennia later, Maas had just made a pavilion in Hannover, and he received an answer back. What are the possibilities, what are the options for the market square? There were plenty of ideas... a markethal? An UFO? A stage with a lifted farm? A tower, as ambitious as the Provincial Parliament in Hertogenbosch?

市场内部的整体性（2002—2004）
Whole in the market (2002~2004)

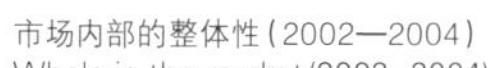

Suddenly, there was a Maecenas in Schijndel that wanted to offer the village a theater. Why not on the market? Too high? Then we just make a whole in the market, with a glass roof so the theater illuminates the sky. In what appeared a local soap, the idea was narrowly voted down in town council. Still no theater.

舞台（2004）
The stage (2004)

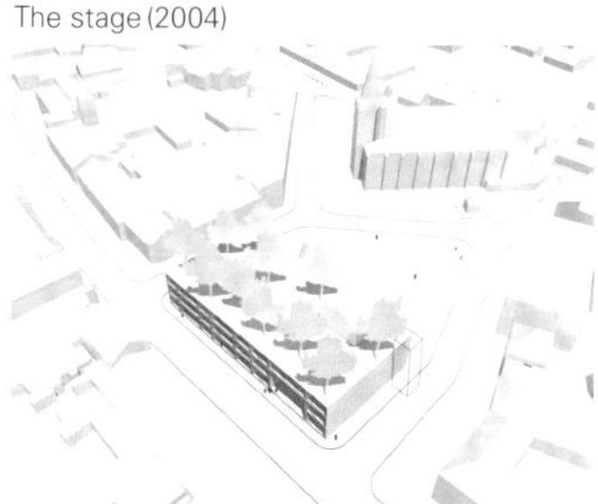

To avoid that the square would stay empty now they've made it a car-free zone, Maas was asked to propose an alternative that would use the maximum volume. There appeared a market hall, an UFO and a stage with a lifted farm.

最大的外围护结构（2005）
Maximum envelope (2005)

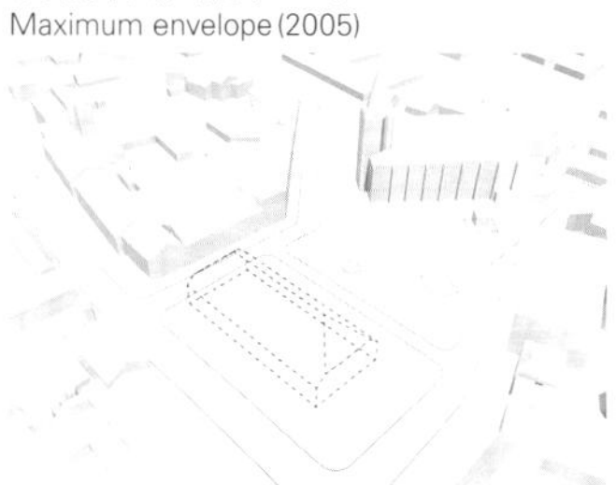

In a committee afterwards, they proposed a maximum envelope for the future development. A bit sturdy, but not too much. Not too short but also not too long. A bit closer to the north side of the square so the old street view would appear. Not too much to the south so there can be a terrace. Not too low or too high for the sloping roofs so there's enough light on the streets. It looks like a farm! A Freudian result.

均衡的农场场地
The average farm

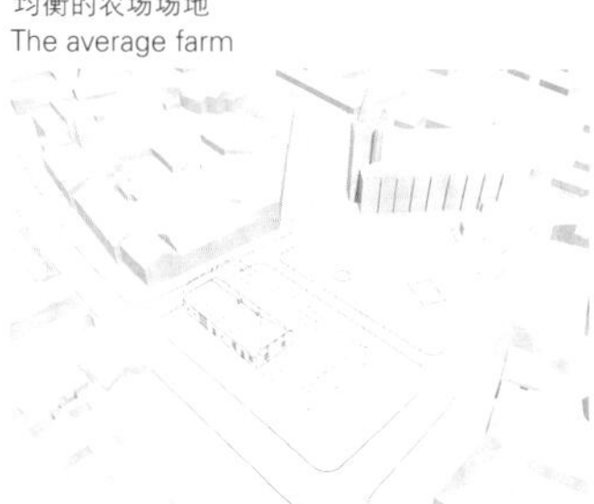

What kind of farm should we make? An exact copy? But of which farm? Why not an average farm? In size and facade. It absorbs the characteristics of all farms in Schijndel. Alright then, only the long facade types. And only the ones with half hipped roofs. And only the ones before the war because that's the era we want to recall. So there are approximately a hundred farms left that come together: average we see 2 windows in the house part and 3 windows in the stable part of the long side. So there are 4 stable doors and 2 chimneys. Etc...

调整为合适的规模（2011）
Scale to fit (2011)

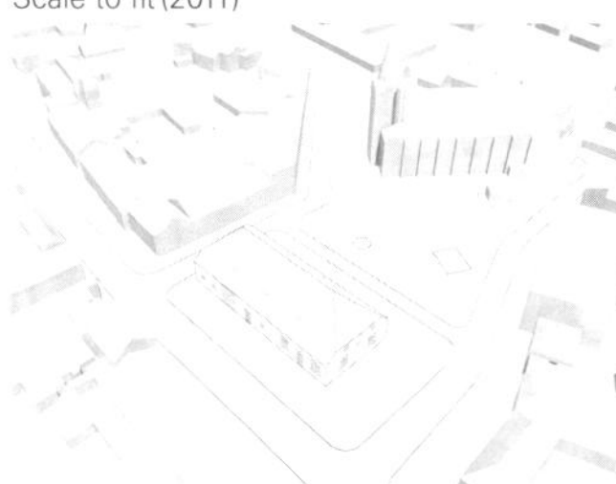

The hard-fought for maximum envelope offers more spaces than the average farm. This means that the average farm could be enlarged by 160 percent. So everything becomes bigger. The doors, the windows, the bricks, the thatch...

玻璃农场（2011）
The Glass Farm (2011)

How do we make that? It shouldn't be an exact copy of a farm. Not even a copy of the average farm. Can't it be a "spirit" of the history? By making the farm out of glass, the "spirit" arises. By printing the average farm on the glass, there arises a contemporary form of stained glass. And that print has more possibilities...

印制（2012）
The print (2012)

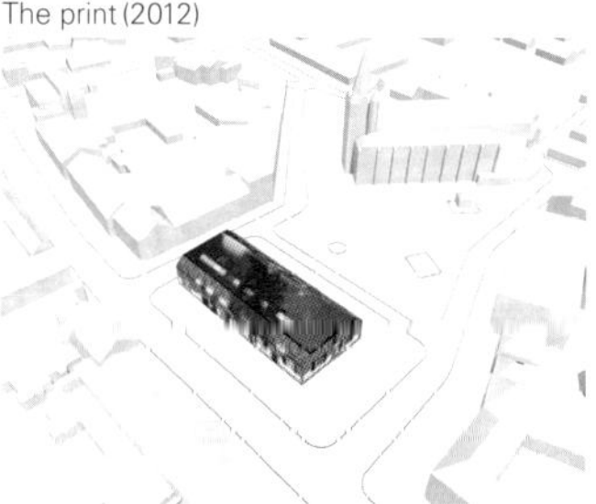

In the windows of the print are several references to the traditional use of the long side farm included. In the reflection of the printed windows, one can see the landscape of Schijndel. By projecting the windows on the opposite wall, the suggestion of transparency is increased. Funny!

[illegible], parts of the prints are erased. So windows appear; showcases for shops, places for advertising, views for the offices appear, too. There are also a shadow of a "lost" tree added, and a reflection of a sunset.

Calimero（2012）
Calimero (2012)

Around the farm, there is a yard. Everything there is also a little bit larger. The hedge, the table, the see-saw. Just like Alice in Wonderland. Does this fit now in the Culture of Brabant to enlarge things like the chair of Oirschot and the lump of Sint Oedenrode? Is this a certain inferiority, a calimero-like attitude? The village became a town.

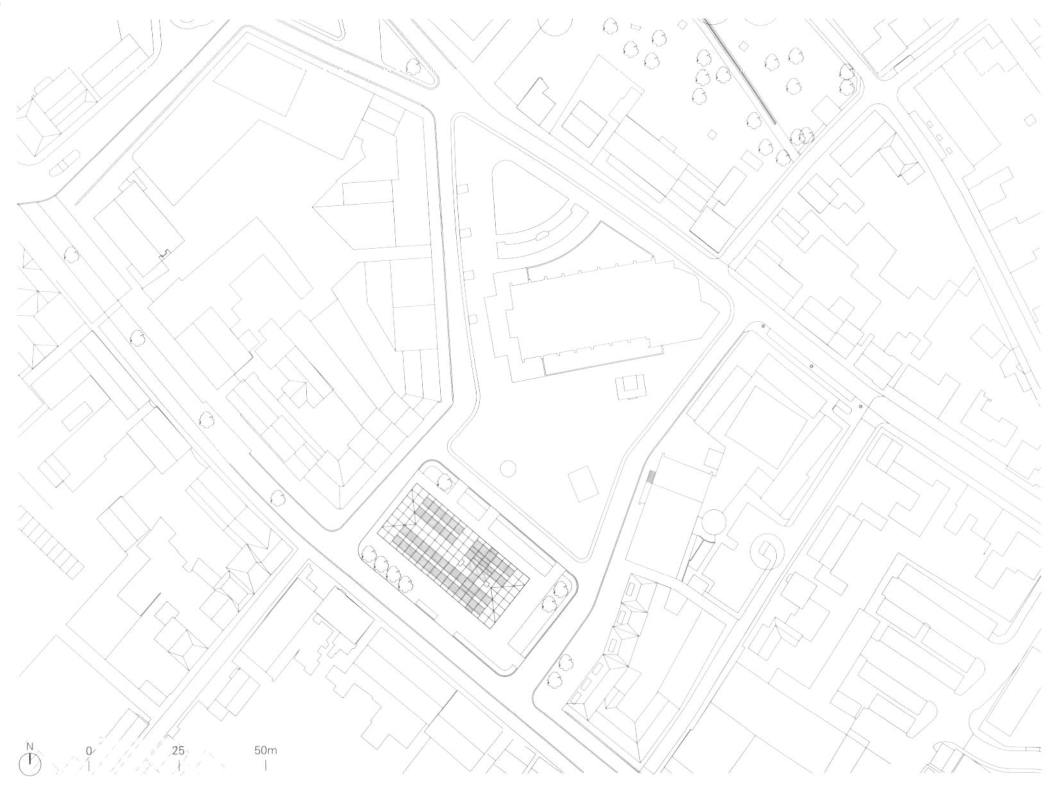
N
0
25
50m

西南立面 south-west elevation

东南立面 south-east elevation

A-A' 剖面图 section A-A'

西北立面 north-west elevation

B-B' 剖面图 section B-B'

C-C' 剖面图 section C-C'

(073)
544 21 00

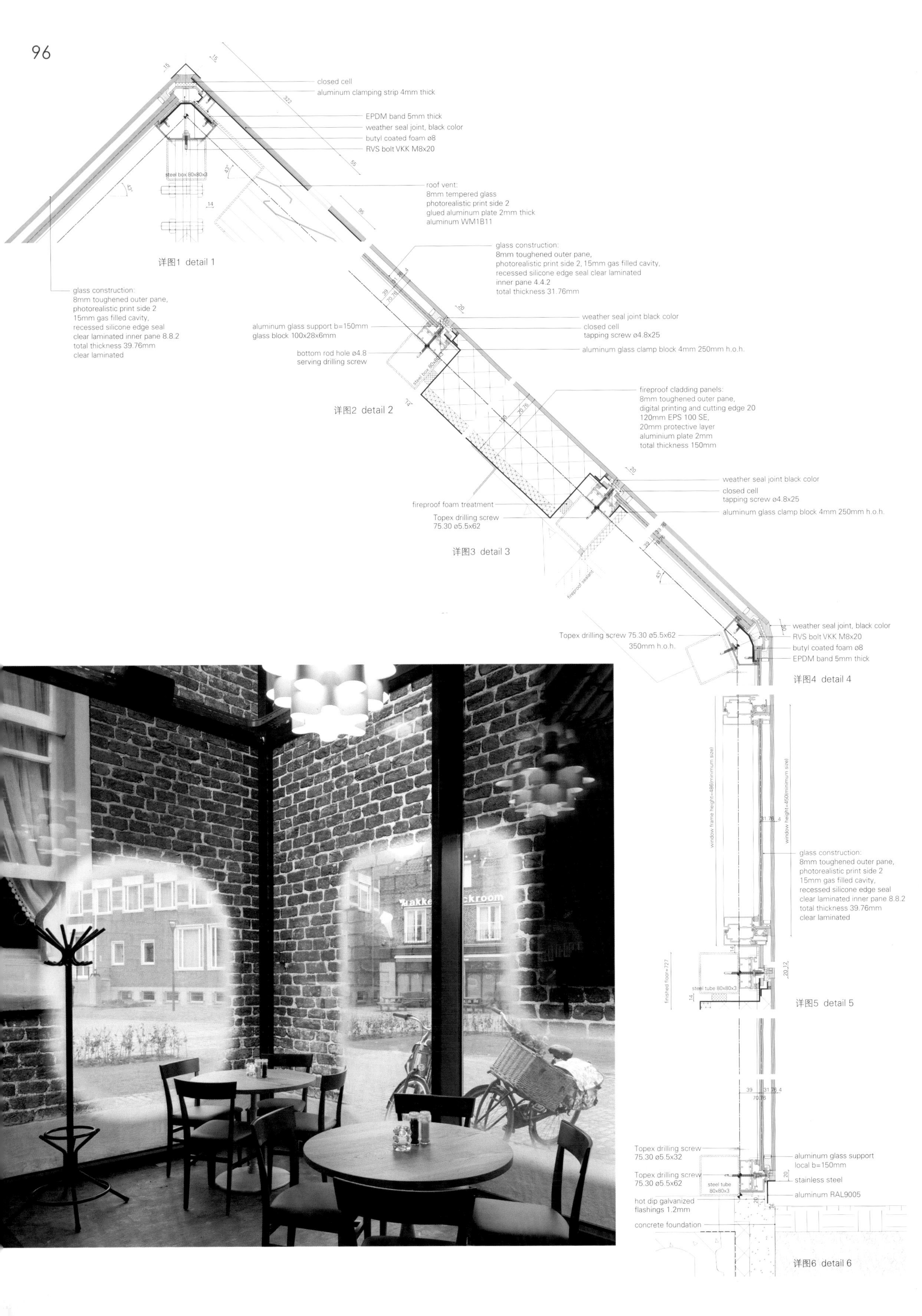

closed cell
aluminum clamping strip 4mm thick
EPDM band 5mm thick
weather seal joint, black color
butyl coated foam ø8
RVS bolt VKK M8x20
steel box 80x80x3
roof vent:
8mm tempered glass
photorealistic print side 2
glued aluminum plate 2mm thick
aluminum WM1B11
详图1 detail 1
glass construction:
8mm toughened outer pane,
photorealistic print side 2
15mm gas filled cavity,
recessed silicone edge seal
clear laminated inner pane 8.8.2
total thickness 39.76mm
clear laminated
glass construction:
8mm toughened outer pane,
photorealistic print side 2, 15mm gas filled cavity,
recessed silicone edge seal clear laminated
inner pane 4.4.2
total thickness 31.76mm
aluminum glass support b=150mm
glass block 100x28x6mm
bottom rod hole ø4.8
serving drilling screw
weather seal joint black color
closed cell
tapping screw ø4.8x25
aluminum glass clamp block 4mm 250mm h.o.h.
详图2 detail 2
fireproof cladding panels:
8mm toughened outer pane,
digital printing and cutting edge 20
120mm EPS 100 SE,
20mm protective layer
aluminium plate 2mm
total thickness 150mm
weather seal joint black color
closed cell
tapping screw ø4.8x25
aluminum glass clamp block 4mm 250mm h.o.h.
fireproof foam treatment
Topex drilling screw
75.30 ø5.5x62
详图3 detail 3
fireproof sealant
Topex drilling screw 75.30 ø5.5x62
350mm h.o.h.
weather seal joint, black color
RVS bolt VKK M8x20
butyl coated foam ø8
EPDM band 5mm thick
详图4 detail 4
window frame height=486(minimum size)
window height=450(minimum size)
glass construction:
8mm toughened outer pane,
photorealistic print side 2
15mm gas filled cavity,
recessed silicone edge seal
clear laminated inner pane 8.8.2
total thickness 39.76mm
clear laminated
finished floor+727
steel tube 80x80x3
详图5 detail 5
Topex drilling screw
75.30 ø5.5x32
Topex drilling screw
75.30 ø5.5x62
steel tube 80x80x3
hot dip galvanized
flashings 1.2mm
concrete foundation
aluminum glass support
local b=150mm
stainless steel
aluminum RAL9005
详图6 detail 6

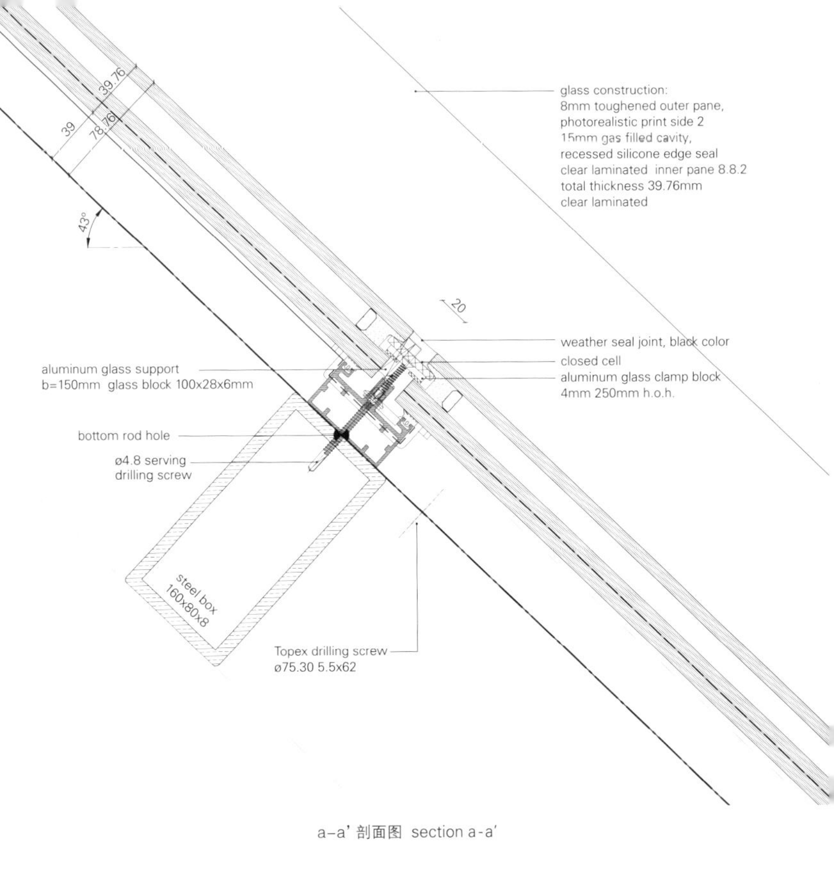

a–a' 剖面图 section a-a'

adversaries. The 1600m² building which is entirely covered by a glass facade consists primarily of a series of public amenities such as restaurants, shops and a wellness center.

By coincidence, the maximum envelope that was defined by the town planners had the form of a traditional Schijndel farm. All remaining historical local farms were measured, analyzed and an "ideal" average was conceived from these data. In collaboration with MVRDV, artist Frank van der Salm photographed all the remaining traditional farms, and from these an image of the "typical farm" was composed. This image was printed using fritted procedure onto the 1800m² glass facade, resulting in an effect such as a stained glass window in a cathedral. The print is more or less translucent depending on the need for light and views.

At night the structure will be illuminated from the inside, becoming a monument to the farm. At a height of 14 meters the Glass Farm is intentionally designed out of scale and is 1.6 times larger than a real farm, symbolizing the village growing into a town. The printed image follows this "augmented history", with the superimposed farm door appearing 4 meters tall. When adults interact with the building, they can experience toddler size again, possibly adding an element of nostalgic remembrance to their reception of the building. To enhance this further, there will be a table and swing next to the building, to become a scaled up farmyard.

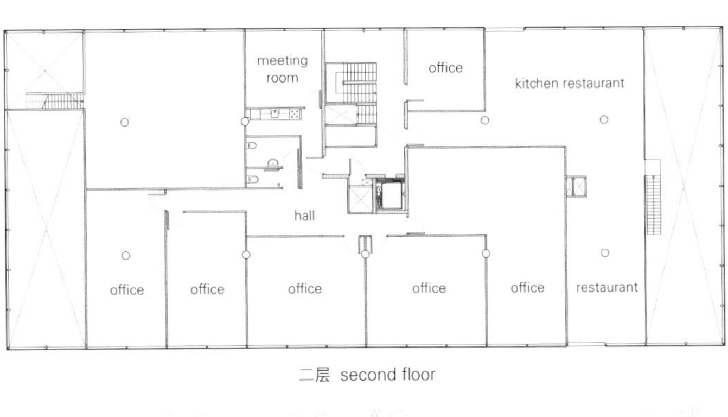

二层 second floor

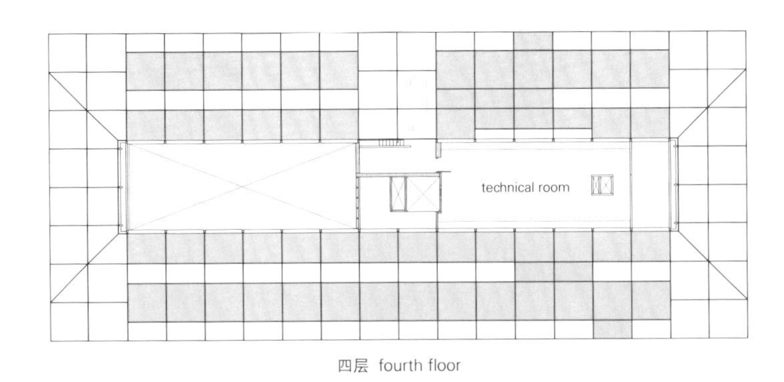

四层 fourth floor

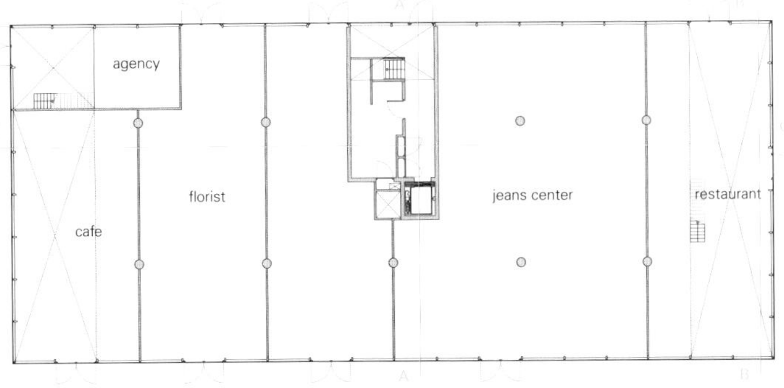

一层 first floor

三层 third floor

0 2 5m

do Morro竞技场

Herzog & de Meuron

麦·路易莎的do Morro竞技体育馆是在纳塔尔市的一个贫穷社区实现获得高标准的社会设备的梦想，能够极大地为居民更美好的未来做出贡献。

分享这个梦想的赞助人是麦·路易莎的Centro Sócio-Pastoral Nossa Senhora da Conceição(CSPNSC)和亚美洛巴基金会（一个瑞士的发展援助机构）。关于在街区创造文化和体育空间的地方需求的第一次对话于社区讲座期间开始，由CSPNSC在2006年推动，以讨论地方发展。

该项目是大众积极性所产生的结果，引领了州政府、纳塔尔市政当局、CSPNSC、亚美洛巴基金会、教师、学生、学校工作人员和麦·路易莎社区的伙伴关系。列入这方面的经验独立于政治上的分歧、政党和宗教信仰，表明一个伟大的项目是可能超越一切个人利益的，为了项目的成功是可以克服障碍和困难的，并使所有人受益。

do Morro竞技体育馆是亚美洛巴基金会捐赠给麦·路易莎社区的，因为它位于公立学校内，也是捐赠给州政府的，而州政府通过Dinarte Mariz学校，能够利用体育馆设施来进行运动和体能教育活动。

该项目的另一个显著后果是州政府决定把Dinarte Mariz学校转为技校。而第二个阶段现在已开始，将以新技校、体育馆、CSPNSC之间的合作关系为特点，致力于对邻里年轻一代影响深远的社交空间的建造。

这一充满着美感和极具功能性的项目是由瑞士的赫尔佐格&德梅隆建筑事务所没有花费任何成本研发出来的。它也是一个整个城市都为之骄傲的建筑标志。

赫尔佐格&德梅隆事务所也创造了"麦·路易莎的视野"，它是一个丰富的城市规划项目，提出一系列的街区嵌入结构，位于Alameda Sabino Gentili、由纳塔尔市政局进行建设的一条长长的步行街，即格林大街之间。

该体育馆是"视野"的一部分，将促进年轻人花费时间做运动，进行文化和休闲活动，帮助他们离开街头。此外，它也将为老年人、超重人群、有特殊需要的人以及整个社区参与体育活动提供场地。目前的挑战是使体育馆的工作充分发挥其潜力，形成以社区为导向的议程。

该项目将成为社会独立和发展的真正工具。为了巩固活动的议程和启发其他社区相似的公共政策，它将推动公众和私人伙伴之间的良好关系。

1 沙丘中的广场
2 工作室
3 体育场
4 商店
5 体育活动和舞蹈场地
（do Morro竞技场）
6 学校
7 全新的连接区
8 文化区
9 公园
10 海边步道
11 绿色天篷
1. plaza at the Dunes
2. workshop
3. sports field
4. shops
5. sport and dance (Arena do Morro)
6. school
7. a new link
8. culture
9. park
10. sea walkway
11. green canopy
N
0
100
300m

Arena do Morro

Mãe Luiza's Arena do Morro gymnasium is the realization of the dream of a poor community from the city of Natal, to have access to high standard social equipment, capable of great contributing to the construction of a more decent future for its inhabitants. The patrons who share this dream are Mãe Luiza's Centro Sócio-Pastoral Nossa Senhora da Conceição(CSPNSC) and the Ameropa Foundation, a Swiss institution for development aid. The first conversation regarding the local need to create spaces for culture and sports in the neighborhood started during a community seminar, promoted by CSPNSC in 2006, which discussed local development.

The project was the result of a common enthusiasm leading to a partnership between the state's government, the municipality of Natal, the CSPNSC, the Ameropa Foundation, teachers, students, school's staff, and Mãe Luiza's community. This experience of inclusion, independent from political differences, party and religious beliefs, showed that it is possible, beyond all private interests, and together, to overcome obstacles and difficulties for the success of a great project that will benefit all.

项目名称：Arena do Morro
地点：Rua Camaragibe, Mãe Luiza, Natal, Brazil
建筑师：Herzog & de Meuron
项目合作者：Partner in charge_Jacques Herzog, Pierre de Meuron, Ascan Mergenthaler/Markus Widmer
项目团队：Associate, Project director_Tomislav Dushanov/Project manager_Mariana Vilela/Digital technologies_Melissa Shin, Diogo Rabaça Figueiredo, Kai Strehlke, Edyta Augustynowicz, Daniel Fernández Florez
设计顾问：Herzog & de Meuron
执行建筑师/电气工程师/管道工程师/结构工程师/可持续性顾问/景观设计师：PLANTAE-Planejamento Técnico em Arquitetura e Engenharia LTDA.
照明顾问：Luminárias Projeto / 交通顾问：RITUR
功能：sports, recreation
用地面积：5,207m² / 总建筑面积：1,861m² / 有效楼层面积：1,964m²
设计时间：2011 / 施工时间：2012 / 竣工时间：2014
摄影师：
©Iwan Baan(courtesy of the architect)-p.102, p.103, p.104~105, p.109, p.112~113
©Leonardo Finotti-p.100~101, p.106~107, p.108, p.111, p.116, p.117

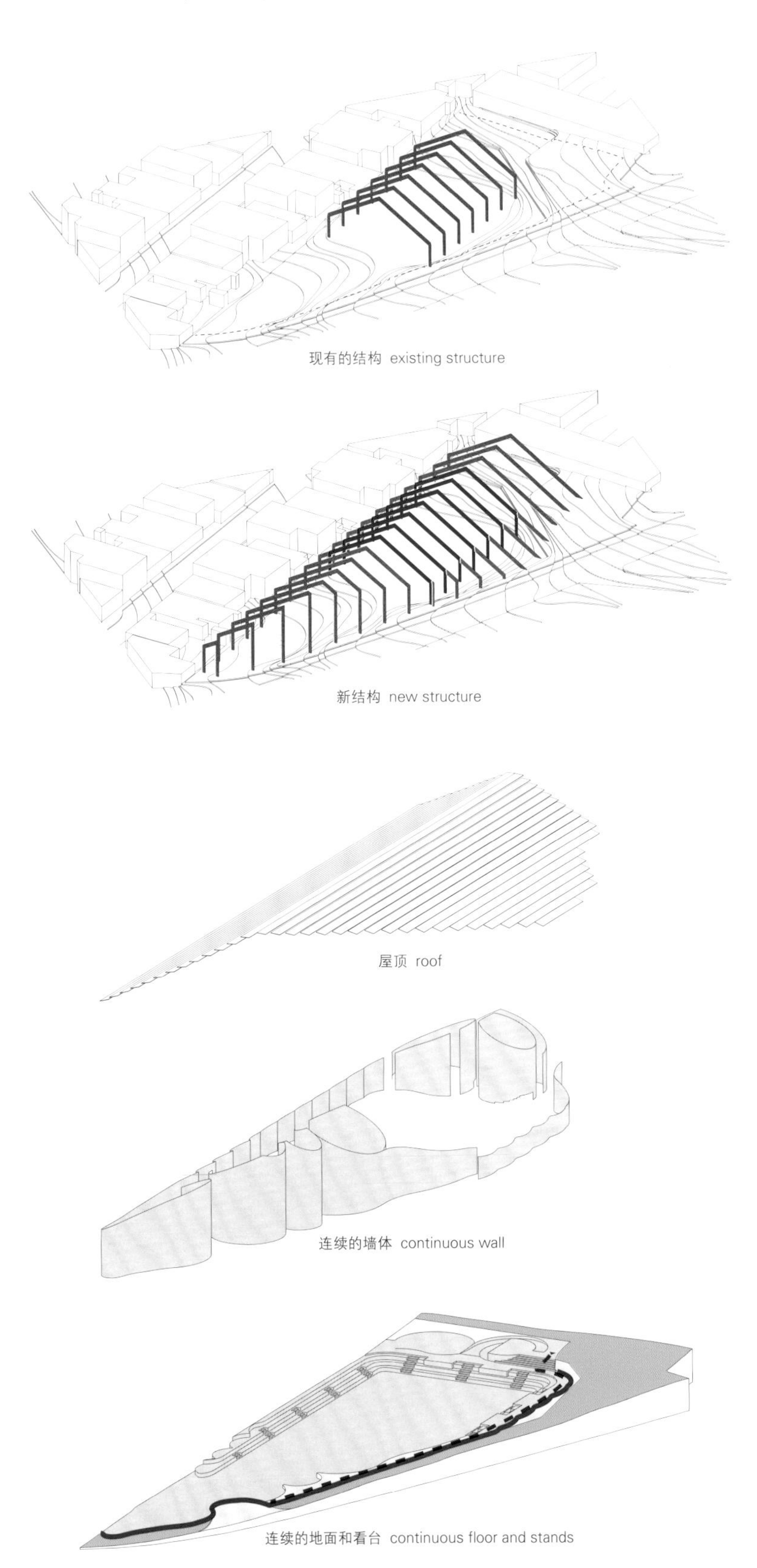

现有的结构 existing structure

新结构 new structure

屋顶 roof

连续的墙体 continuous wall

连续的地面和看台 continuous floor and stands

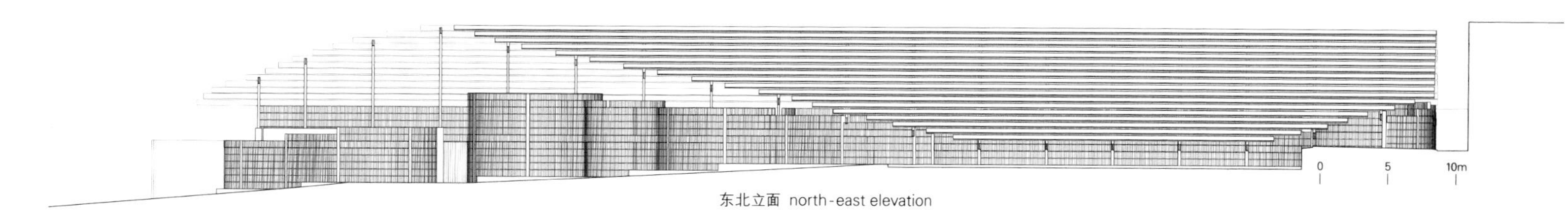

东北立面 north-east elevation

HELP CARR

The Arena do Morro gymnasium is a donation from the Ameropa Foundation to Mãe Luiza's community and also, since it is located in a state's public school, a donation to the state's government, who will be able to, through the school of Dinarte Mariz, use the facilities of the gymnasium for sports and physical education activities.

Another significant consequence of the project was the state government's decision to convert the Dinarte Mariz School into a technical school. The second stage, which begins now, will be marked by the cooperation between the new technical school, the gymnasium, and the CSPNSC, for the construction of far-reaching spaces of social inclusion for the neighborhood's young generation.

This project of great beauty and functionality was developed at no cost by the Swiss architectural practice Herzog & de Meuron. It is also an architectural icon that will honor our city.

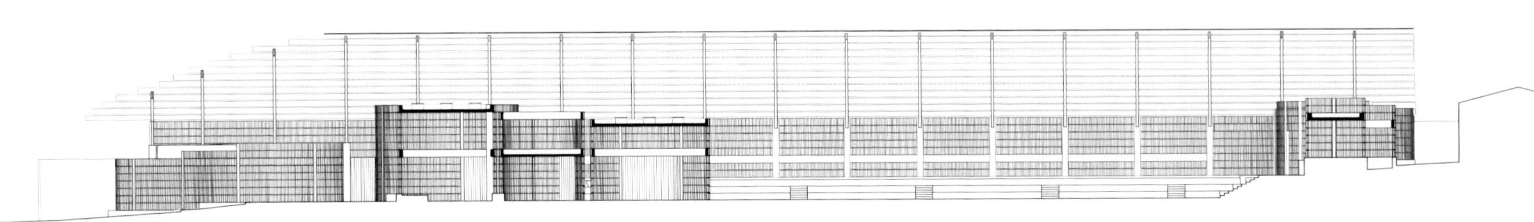

A–A' 剖面图 section A-A'

The practice Herzog & de Meuron also produced the "Vision for Mãe Luiza", a rich urban planning project which suggests a diversity of interventions in the neighborhood, among the Green Street, a long pedestrian street, currently under construction by the municipality of Natal on the Alameda Sabino Gentili.

The gymnasium is part of the "Vision" and will enable young people to spend their time doing sports, culture, and leisure activities, helping to take them out of the streets. In addition, it will also enable sports activities for the elderly, the overweight, people with special needs and for the whole community. The current challenge is to make the gymnasium work to its full potential, enabling this community-orientated agenda.

The project will become a real tool of social independence and development. It will promote rewarding public and private partnerships in order to consolidate the activity's agenda and inspire similar public policies in other communities.

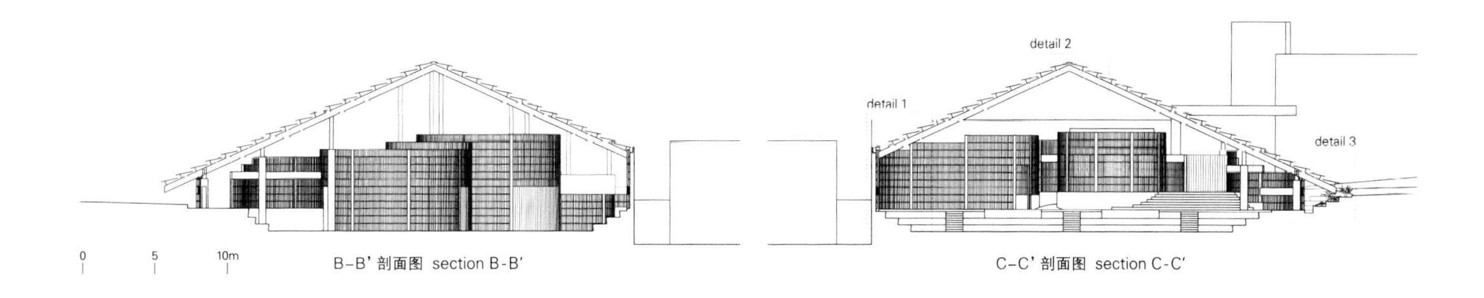

B-B' 剖面图 section B-B'

C-C' 剖面图 section C-C'

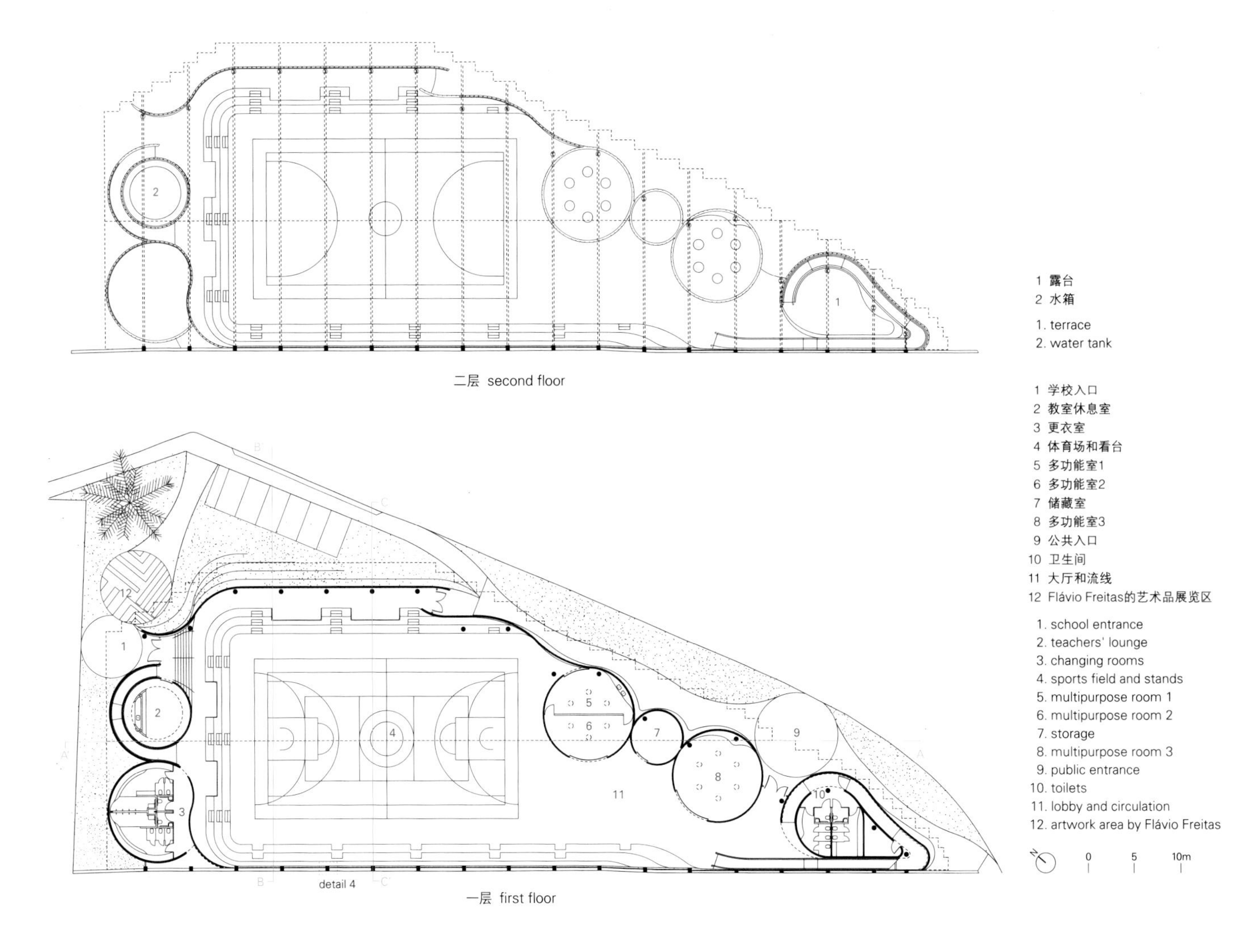
1 露台
2 水箱
1. terrace
2. water tank
1 学校入口
2 教室休息室
3 更衣室
4 体育场和看台
5 多功能室1
6 多功能室2
7 储藏室
8 多功能室3
9 公共入口
10 卫生间
11 大厅和流线
12 Flávio Freitas的艺术品展览区
1. school entrance
2. teachers' lounge
3. changing rooms
4. sports field and stands
5. multipurpose room 1
6. multipurpose room 2
7. storage
8. multipurpose room 3
9. public entrance
10. toilets
11. lobby and circulation
12. artwork area by Flávio Freitas
0
5
10m
detail 4

二层 second floor

一层 first floor

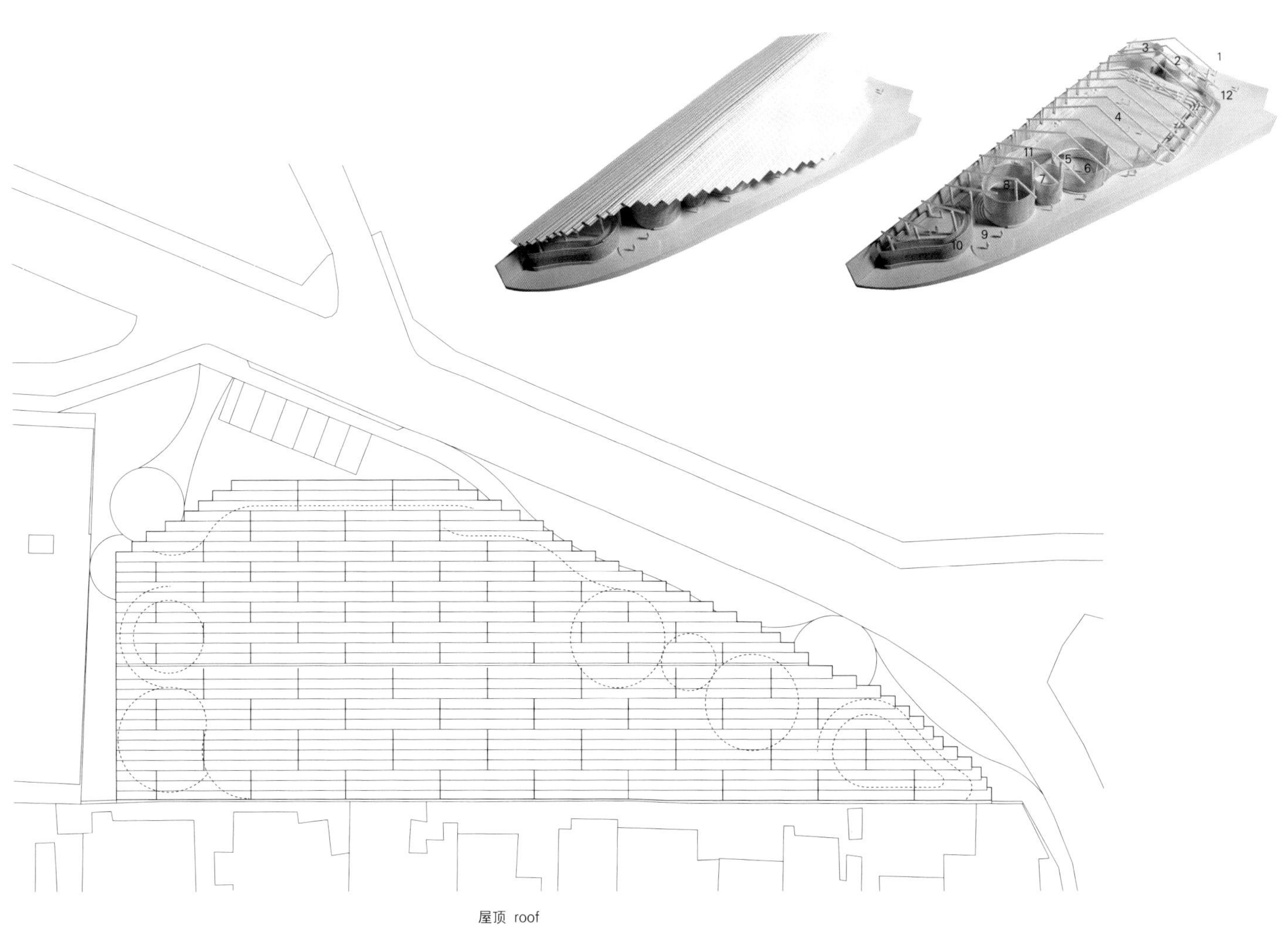

屋顶 roof

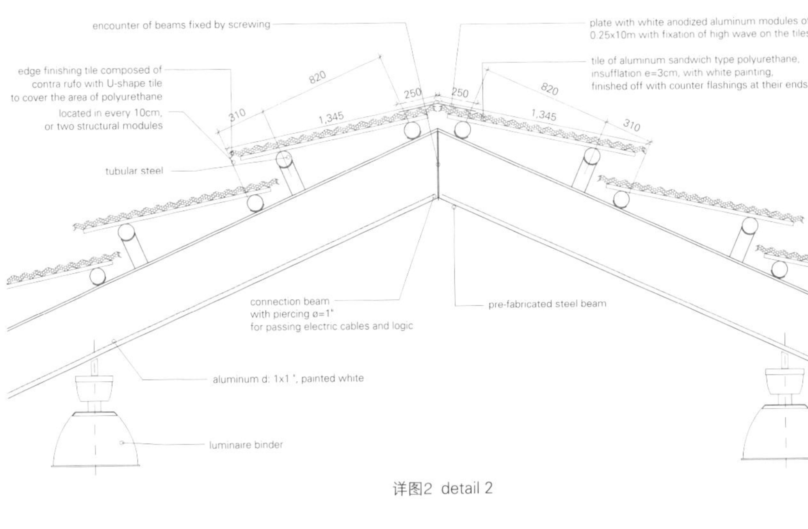

详图2 detail 2

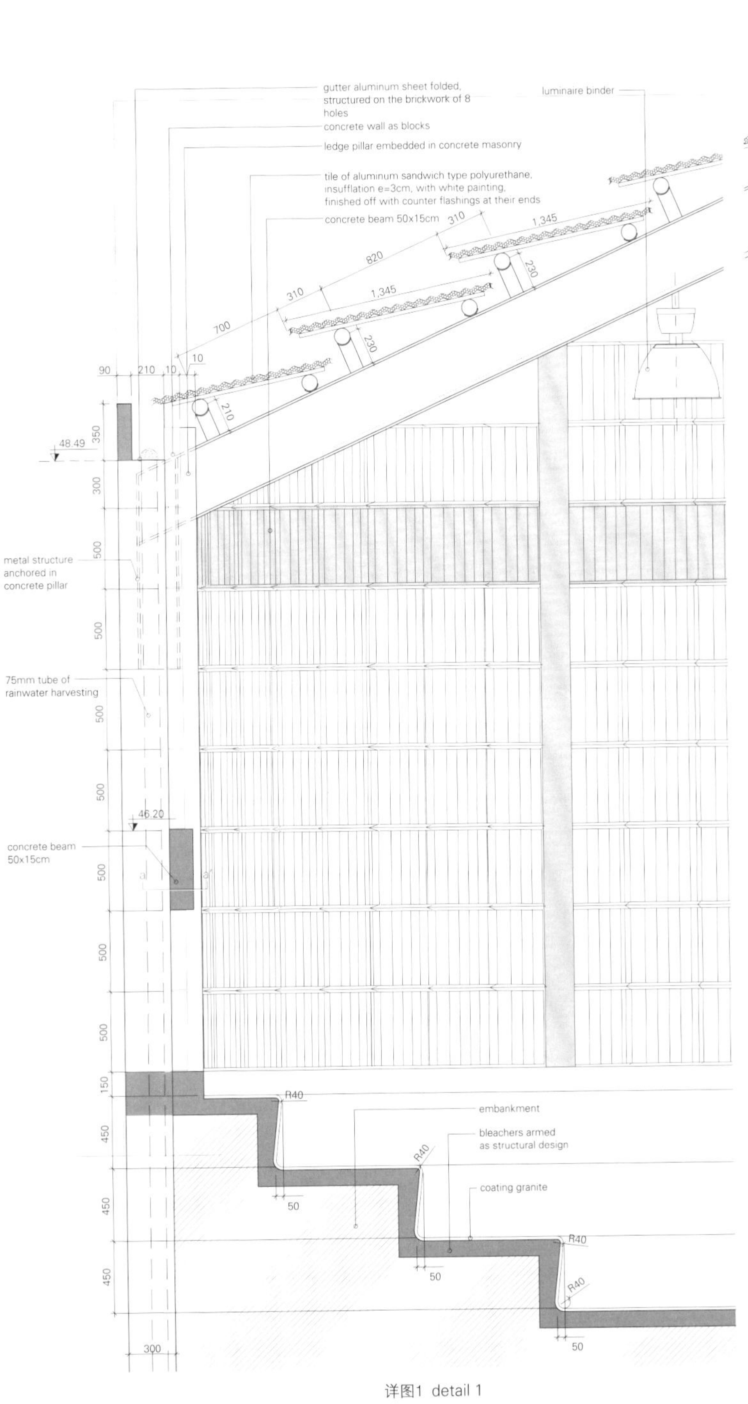

详图1 detail 1

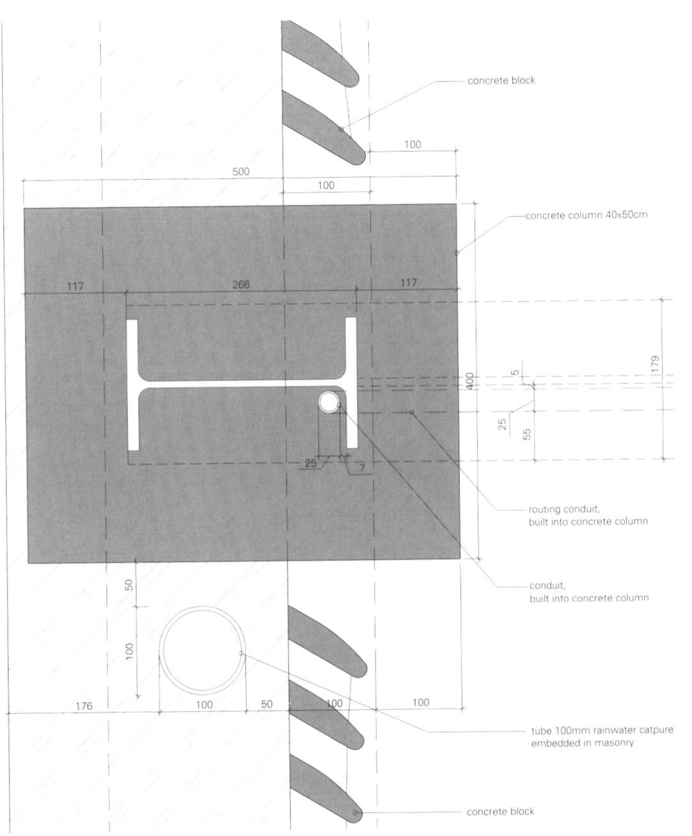

a–a' 剖面图 section a-a'

实体模型中测试的构件
elements to test in the mock up

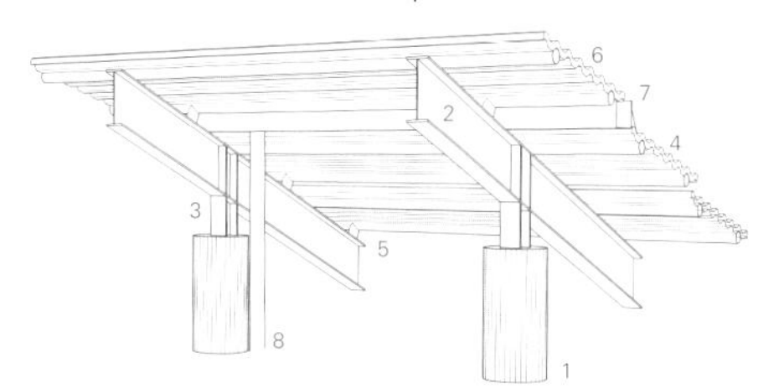

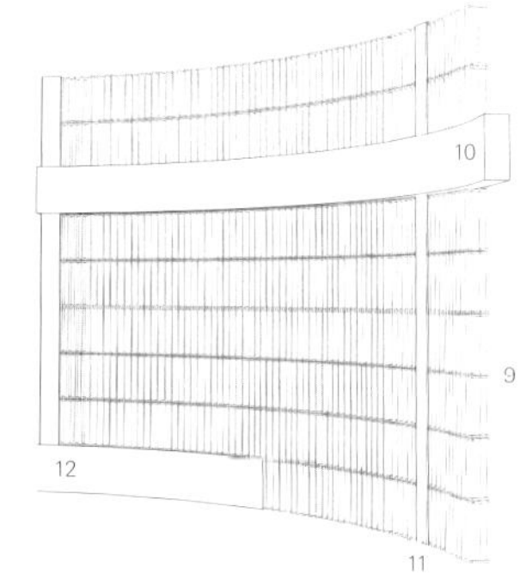

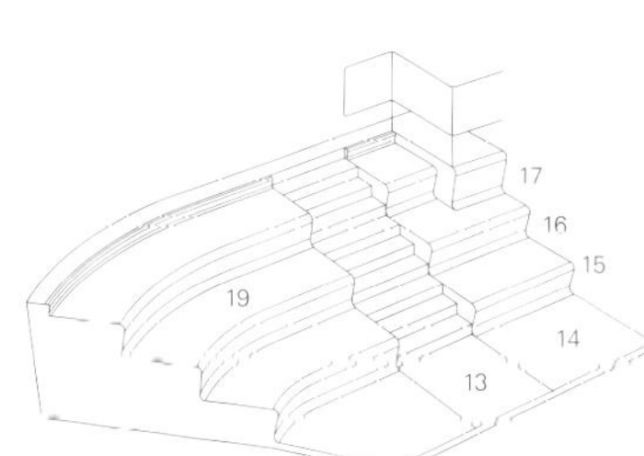

墙体实体模型，2012.4
wall mock up, 2012.4

地面实体模型，2012.4
floor mock up, 2012.4

施工，2013.4
construction, 2013.4

施工，2014.3
construction, 2014.3

施工，2013.9
construction, 2013.9

施工，2014
construction, 2014

1. concrete column ø500mm
2. steel profile "I" W-410X67, 410x179mm
3. steel profile "I" connecting beam column
4. steel rod ø101mm
5. steel rod spacer ø101mm
6. thermoacoustic aluminium panel
7. steel gutter
8. rainwater downpipe ø100mm
9. concrete block
10. concrete beam
11. concrete column
12. base granulite
13. internal flooring-polished granulite
14. external flooring-rough granulite
15. convex radius
16. concave radius
17. joints
18. balustrade
19. curved seating

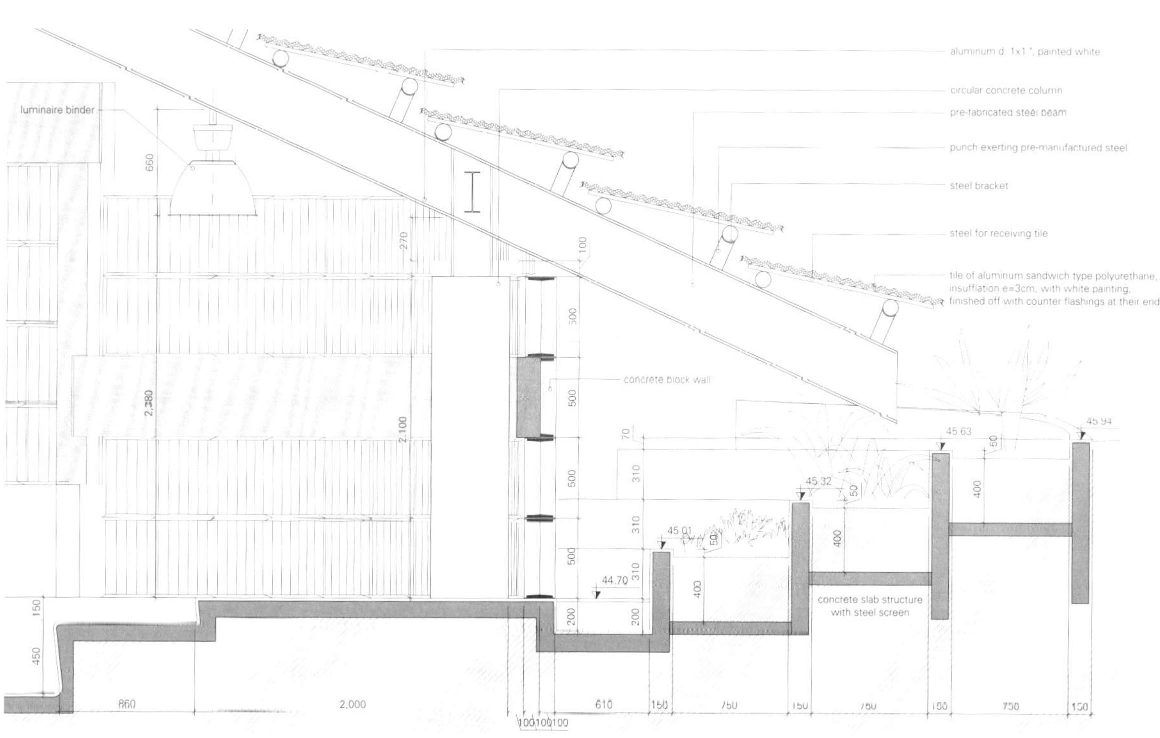

详图3 detail 3

墙砌块类型
wall block type

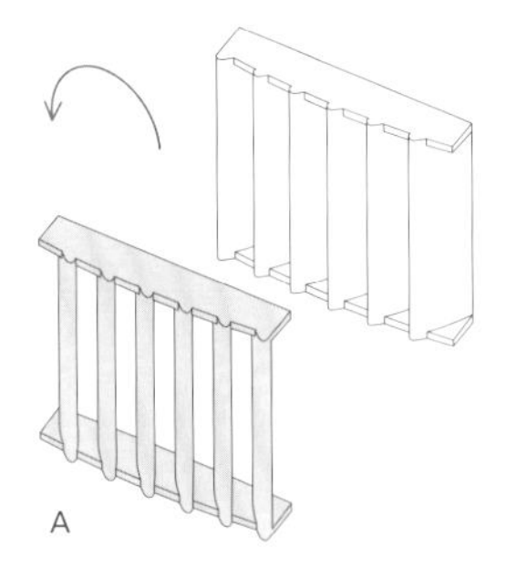

visual permeability
wind
direct sun light
convex shape
54% perforation

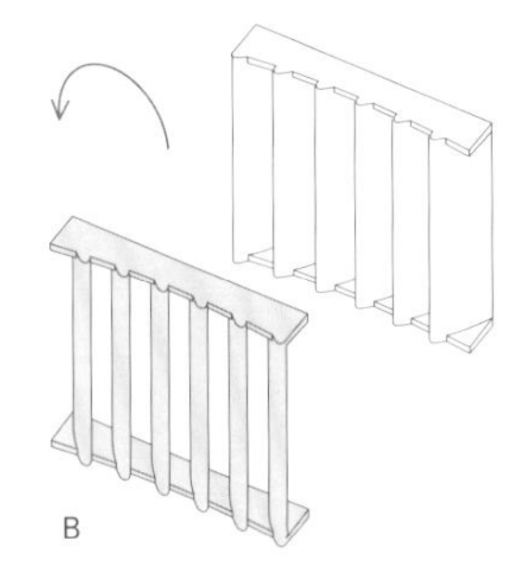

visual permeability
wind
direct sun light
concave shape
54% perforation

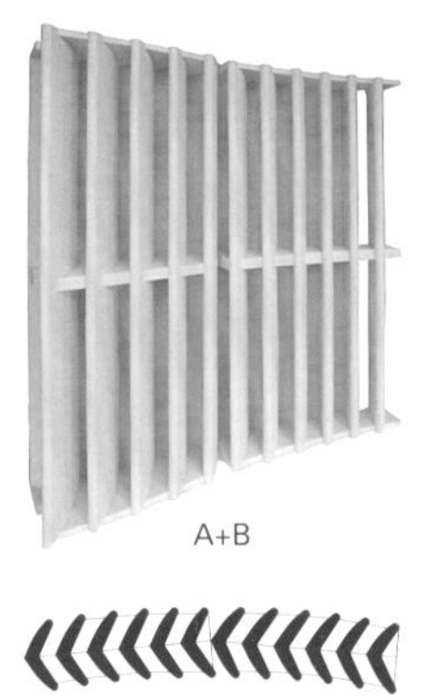

visual privacy
wind
indirect sun light
concave-convex shape

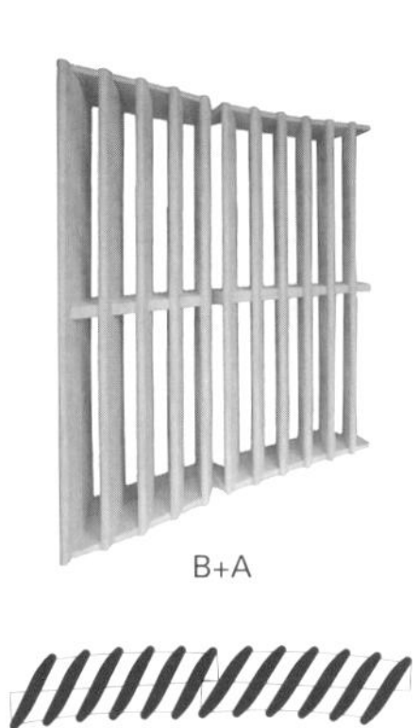

visual permeability
wind
indirect sun light
concave-convex shape

第一个墙砌块类型，2012.3
first block prototype, 2012.3

烘干后的看台，2013.4
drying stand, 2013.4

水上的建筑

Álvaro Siza + Carlos Castanheira

“水上的建筑”在中国淮安市实联化工厂揭幕

这是一个巨大的产业园区。朦胧的烟雾从高耸的烟囱排出。灰白色的建筑物挡在了道路之间的直线上，这些道路间的距离呈量化布局，模仿棋盘的形式。这个地方充满了操作的机器和忙着从一个位置走到另一个位置的工人的沉重声音。这是实联化工厂的样子。

在进入一个工业园区时，人们期望一切都将按照一定的规则和流程进行。然而建筑颠覆了这样的期望，在扩大了面积为100 000m²且深8m的人工湖中间，一座不同性质的建筑宣告着它的存在，倾覆了化工厂综合体内部的现有秩序。

2008年，于台湾玻璃集团庆祝其成立50周年的前几年，总裁林伯实，也是实联集团的主席，设想在人工湖（和水库）上面建造一座办公楼，用于大型工业设施。他希望建筑师是国际知名的，但是对于亚洲来说却是新星，在生活和工作中有广泛而丰富的经验。四位预想的建筑师均是由维特拉设计博物馆的创始合伙人之一，同时兼任布瓦布榭庄园理事的亚历山大·冯·维格萨克推荐的。

在审查建筑师候选人时，林主席决定邀请国际知名的葡萄牙建筑师阿尔瓦罗·西扎来领导设计团队。因此，西扎在中国的第一个项目，“水上的建筑”就这样开始了，随后历时四年的设计和建造。揭幕典礼于2014年8月30日举行。

阿尔瓦罗·西扎在中国的第一座建筑物

位于江苏省淮安市的新盐工业园，江苏实联化工有限公司是世界上最大的纯碱和氯化铵联合生产厂之一。一站式的生产有可能在化工厂进行：该操作包括混合、加工、包装、贮存和运输。所有需要的电力是自身生产的，且工厂声称拥有中国工厂中规模最大且最高级的设施。

鉴于办公楼位于靠近2000m²的工厂综合体的入口处，并代表公司形象，因此项目本身对台湾玻璃企业集团也是极其重要的。

台湾玻璃企业集团邀请了很多人参加建筑物的落成典礼，包括大陆和台湾的建筑师，以及教授、编辑和记者。韩国的C3出版公社也被邀请参加为期三天的落成规划，其议程包括游览建筑内部和参加就职典礼。

第一天的建筑之旅从实联厂的南大门开始。我们通过L形混凝土结构，横跨人工湖上的桥，抵达由阿尔瓦罗·西扎设计的大楼，这座大楼看起来好像是浮在水上的。即使西扎有着60年的建筑经验，但是水上施工对他来说也是一个全新的挑战。

优雅地栖息于水面上的龙

第一次看到“水上的建筑”是通过照片来看的。曲线外形优雅，蜿蜒，且从建筑内散发出的空间感确实势不可挡。曲线形建筑本身的规模是相当巨大的，经测量，长度超过300m，有效楼层面积为11 000m²。

与此同时，站在办公楼前，用肉眼看它，建筑呈现了优雅的气质和风度。这是一座白色的混凝土建筑，线、面无多余组成。两层建筑大面积地匍匐在水面上。它没有设计得比厂区的其他建筑物高很多，使其不会突显得更高。然而，西扎通过赋予其灵活性，来增添一种对建筑的独特触感，使其成为工厂建筑的网格布局上的一个阶梯，这种低调的奢华是西扎设计的核心。

该建筑呈U形或马蹄形，因此建筑物内部的曲线环绕着水。事实上，建筑采用了龙的外形，龙是中国的象征。据西扎所说，建筑包含了他最初的草图内绘制的直线，但随着时间的推移，他决定创造一处空间和流线，使人们之间的沟通顺畅。

其结果是直线变成曲线。后来，有人提到设计成类似龙的造型，之后这成为该建筑的重要主题。

凝望建筑，走进建筑，感受建筑

坐落在水中间的建筑物通过两座桥与陆地连接起来。东面的桥体作为从外部进入建筑综合体的主入口，而北部的桥梁是办公楼和工厂之间的活动入口。

在进入建筑物时，人们第一眼看见的便是信息台和其后方的窗户。窗户为人们框出了水花四溅的场景。入口大厅将建筑的流线进行了划分。左侧是办公场所，右侧则是服务区。

从入门到办公室，走廊陡峭的曲线使人感觉这里好像是没有尽头的空间。这感觉就像被吸了进去。在这处空间里，一切都是白色的，柱子、墙面，甚至是家具。它是安静的、祥和的，甚至连呼吸也听不到。沿着弯曲的走廊前行，穿过窗户向外看，对我们的眼睛来说，茫茫无尽的水花是一种乐趣。上空空间延伸到二层，使自然光照亮室内。

在通往办公场所周围的半路上，有一个小型大堂。在大堂的左侧，会议室伸向水面。这处空间也没有多余的空间来转移分散我们的注意力。在西扎设计的所有室内空间内，会议室天花板的高度有所区分，以便间接照明能够巧妙地照亮室内。会议室的一些显著特色是桌子、椅子和储存家具，这些都是西扎自己设计的。黑褐色的木质家具在完全的白色空间中显得气势十分宏伟。

我们现在回头，朝主入口方向走去，来参观大楼的对面。服务空间的氛围和办公空间的氛围有很大不同。与办公空间的明亮氛围不同，其走廊散发着某种沉闷且柔和的气氛，没有明亮的灯光或窗户。

主入口右侧的一个缓缓倾斜的坡道通往命名为"龙之翼"的礼堂。这是一个五边形的空间，其中的几何形状的墙体呈现略凹的状态。天花板看起来好像是已经被切出三角形形状的薄片，且三角形的一边略微缩进。在背面墙，有一个长长的矩形窗口，其高度比人的腰部略低。这样的设计可以让一个人享受漂浮在水上的感觉。

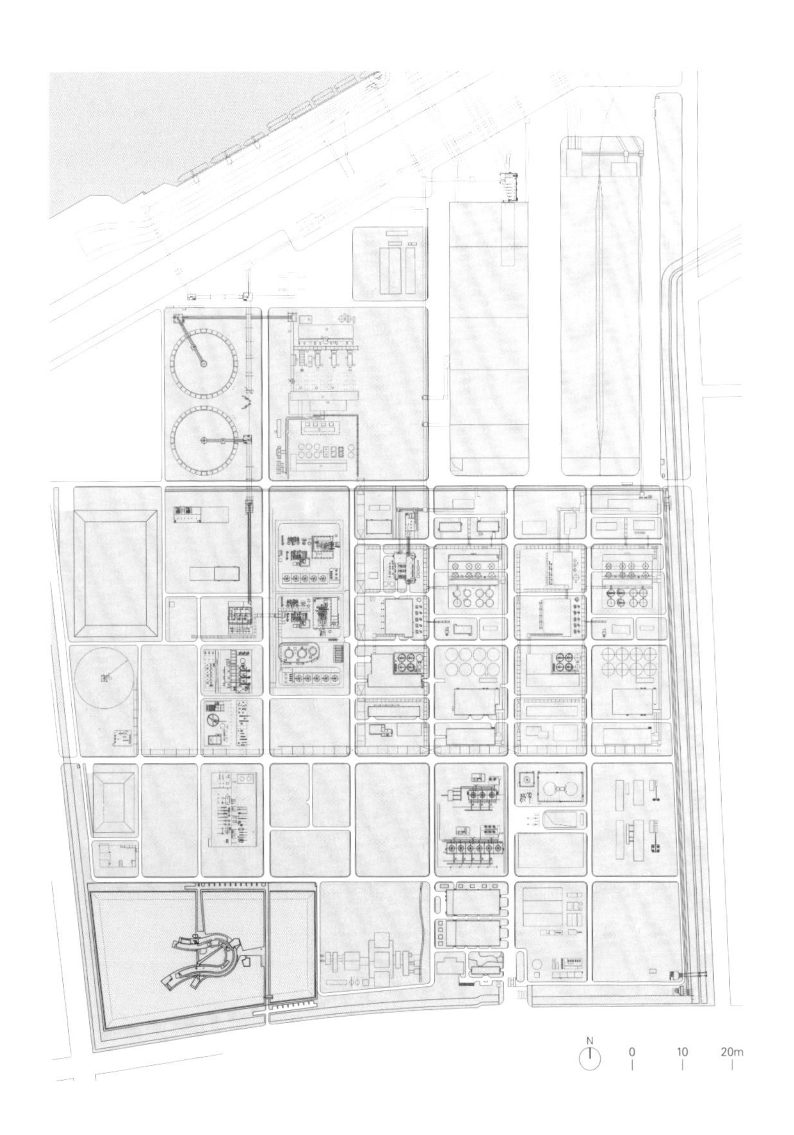

主席办公室、秘书办公室、技术领域和露台位于二楼。氛围比一楼更加明亮、开放。自然而然的，比起内部的空间，人们更重视外面的景色。厂房离办公室很远，因此视角全方位打开，并且没有任何阻碍。

还有二楼的U形建筑物的内侧设有阳台，让人们可以走到外面并观看不同角度的风景。二层的外面进行了装饰，而里面没有。

我们通过斜坡下到一楼。坡道很长，但是坡度较缓。矩形天窗贯穿斜坡的开始和结尾部分，而椭圆形的窗口位于左边和右边。这些设计都让人们享受到户外场景，走在斜坡内部的同时还能够享受到自然光。这个空间在直线和曲线之间以及空白天花板和间接照明之间产生了良好的平衡，这正是西扎喜欢应用的元素。

"水上的建筑"，考虑它的意义

很多人形容阿尔瓦罗·西扎是最有能力通过建筑表达人的感情的人。他通过将物体的物理性能最大化，来把生活融入一处简单的空间。他也在控制穿过建筑的光线的方面表现出色。西扎的杰出才华充分体现在"水上的建筑"中。阿尔瓦罗·西扎的合作伙伴，建筑师Carlos Castanheira说："阿尔瓦罗·西扎在他想创造的雕塑、绘画和写作上有一些文艺复兴时期的男人的特质。这将对美的探索展示为一个他对空间的控制，以及赋予临时空间的一种形式的补充"。

正如林伯实主席所说的那样，"西扎先生始终坚持人与环境之间建立和谐的设计理念。该建筑巧妙地融合了水景和工厂。作为阿尔瓦罗·西扎在中国的第一个项目，这将是未来在中国设计工业厂房的灵感。

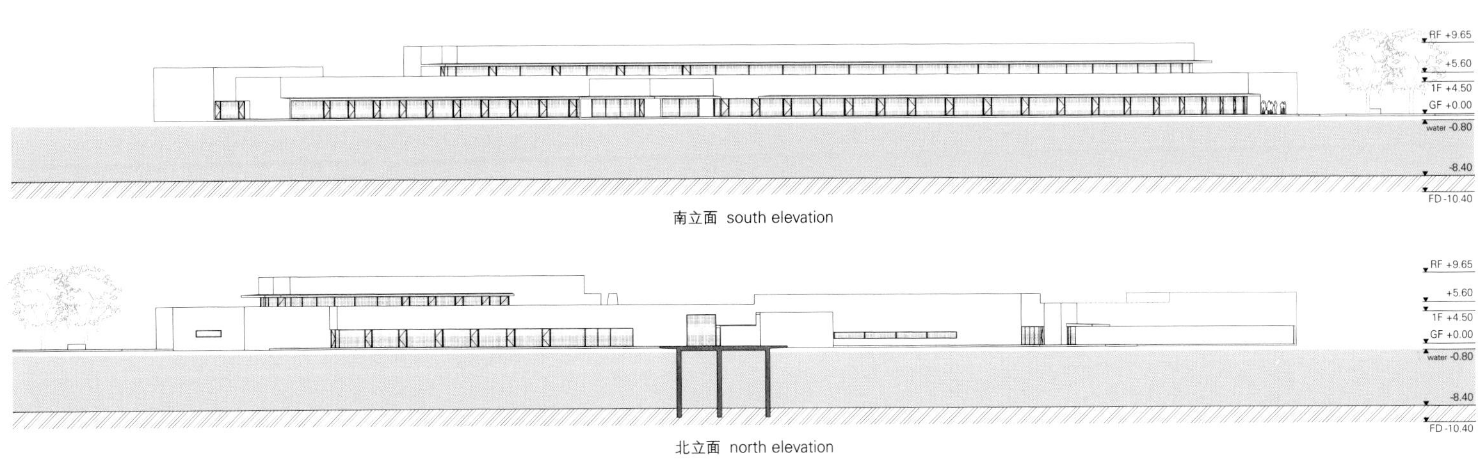

南立面 south elevation

北立面 north elevation

台湾成功大学的Shuenn–Ren Liou教授指出："水上的建筑"将是中国建筑社会的一个转折点。他补充说，中国需要放慢脚步，深吸一口气，因为他希望这座由阿尔瓦罗·西扎设计的建筑能够作为设计其他建筑的动力，引导人们去思考空间的本质。

对"水上的建筑"这样的积极评价并不仅限于上面提到的那些。许多人对其进行赞美，是因为其"存在价值"的意义，而不是建筑细节。中国一度被戏称为建筑实验室。以2008年北京奥运会为起点，大部分中国的建筑社会已经热衷于地标性建筑，以提供视觉上的享受。也许，许多对西扎项目的积极评价是希望建筑师能够克服目前的局限性，并开始认同不同维度的价值。

The Building on the Water

"The Building on the Water" unveiled at Shihlien Chemical Plant in Huai'an City, China

It is a huge industrial park. Hazy smoke blows from high-rise chimneys. Achromatic buildings stand in straight lines between roads that are arranged in quantified distances, mimicking a checkerboard. The place is full of the heavy sounds of machines operating and of workers busily walking from one location to another. This is what the Shihlien Chemical Plant looks like.

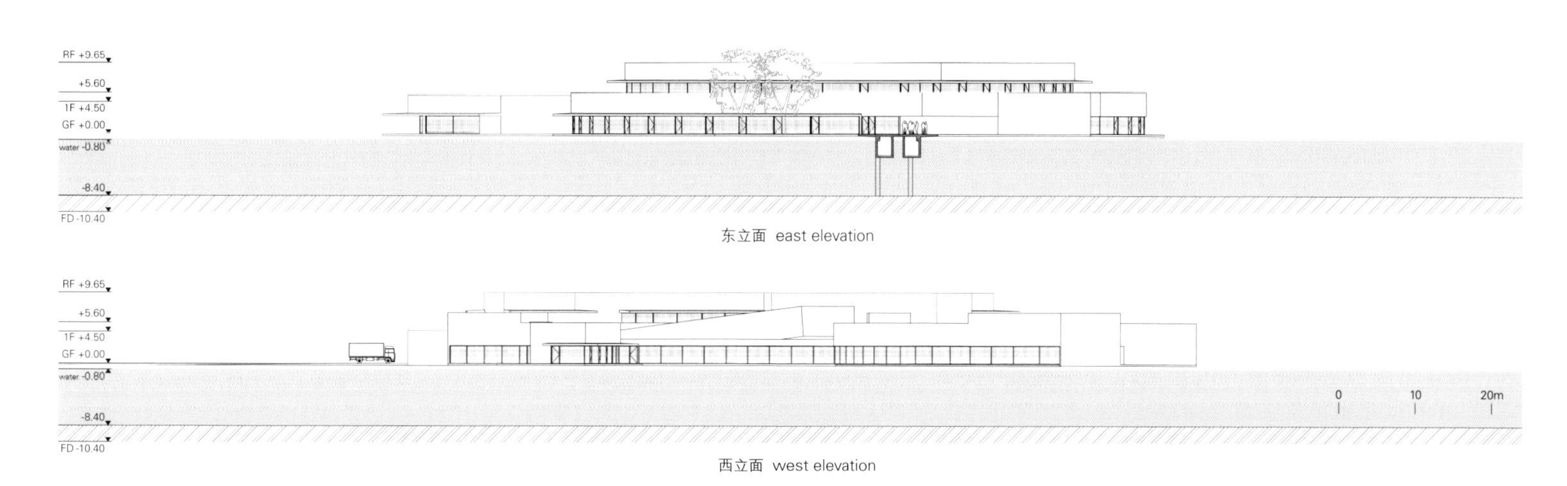

东立面 east elevation

西立面 west elevation

Upon entering an industrial park, one expects that everything will move according to certain rules and processes. Subverting to such an expectation, in the middle of the artificial lake expanding area of 100,000 m² and depth of 8 m, a building of a different nature declares its presence, overturning the existing order within the chemical plant complex.

In 2008, with several years remaining before the Taiwan Glass Group celebrated its 50th anniversary, President Por-Shih Lin who is also the Chairman of Shihlien Group, envisioned creating an office building over the artificial lake (and water reservoir) of the vast industrial complex. He wanted an architect who was internationally renowned but new to Asia, with extensive and rich experience in life and work. Four prospective architects were recommended by Alexander von Vegesack, one of founding partners of the Vitra Design Museum and director of Domaine de Boisbuchet.

Upon reviewing the candidate architects, Chairman Lin decided to invite the internationally renowned Portuguese architect Álvaro Siza to head the design team. Thus, Siza's first project in China, "The Building on the Water" commenced, which subsequently took four years to design and construct. The inauguration was held on August 30th, 2014.

Álvaro Siza's first building in China

Located in the New Salt Industrial Park of Huai'an City, Jiangsu Province, Shihlien Chemical Industrial Jiangsu Co. is one of the world's largest combined soda ash and ammonium chloride production plants. A one-stop manufacturing is possible at the chemical plant: the operations include blending, processing, packaging, storage and shipping. All the electricity required is self-generated, and the plant claims the largest-scale and highest-

level of facilities in China.

Given that the office building is to be located near the entrance of the 2km² plant complex, and will represent the company's image, the project itself was immensely important to the Taiwan Glass Group.

The Taiwan Glass Group invited many people to the building's inauguration ceremony including architects in mainland and Taiwan of China, as well as professors, editors and journalists. In Korea, C3 was invited to take part in the three-day inaugural program, the agenda of which included a building tour and an inaugural ceremony.

The tour of the building on the first day started from the southern gate of the Shihlien Plant. We passed the L-shaped concrete structure, crossed the bridge on the artificial lake and arrived at the building designed by Álvaro Siza, which looks as if it is floating on water. Even with 60 years of experience in architecture, over-water construction was an entirely new venture for Siza.

A dragon perched elegantly on water

The first visual encounter with "The Building on the Water" was through photographs. The shape of curves meandered in a grand manner, and the sense of space that radiates from the building was indeed overwhelming. The curvilinear building itself is quite huge, measuring over 300 meters in length with a total floor area of 11,000 m².

Meanwhile, standing in front of the office building and seeing it with the naked eye, the building had an aura of elegance and grace. It is a white concrete building composed of lines and planes without superfluity. The 2 story building sprawled widely

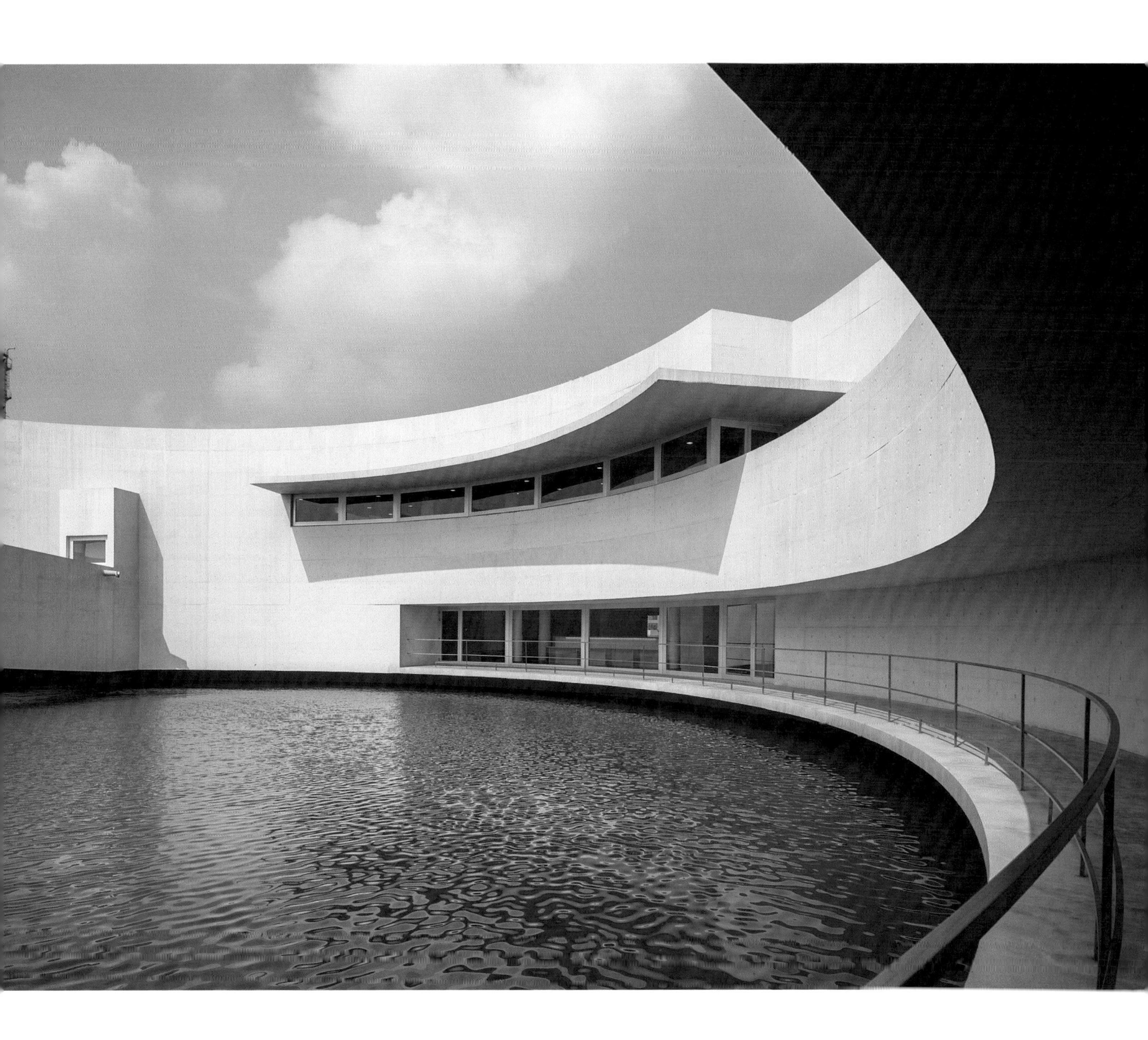

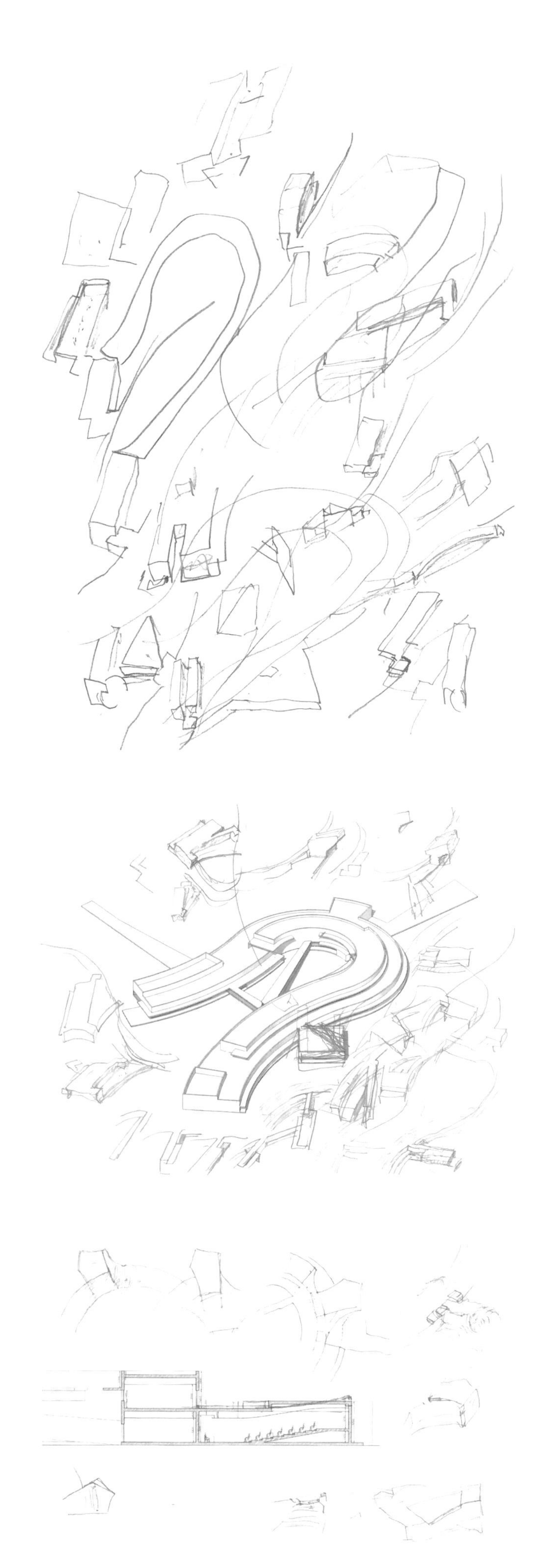

over the water. It was designed not to be excessively higher than other buildings in the plant complex in order not to stand out. Still, Siza added a touch of uniqueness to the building by shaping it with a flexibility, allowing it to be a step above the organized grid arrangement of the plant complex. This subtle flourish is the core of Siza's design.

The building is U-shaped or horseshoe-shaped, hence the curvilinear inner part of the building embraces water. In fact, the building takes the form of a dragon, the symbol of China. According to Siza, the building comprised straight lines in his initial sketch, but over time, he decided to create a space and flow of path which enable smooth communication among people.

As a result, straight lines became curved ones. Later, it was mentioned that the design resembled a dragon, which then became an important motif for the building.

Looking at, walking in and getting a sense for the building

The building sitting in the middle of the water attaches to land via two bridges. The eastern bridge serves as the main entrance for entering the complex from outside, and the northern bridge serves as an entrance for moving between the office building and the plant complex.

The first thing that people see when entering the building is the information desk and the window behind it which frames a scene of splashing water. The entrance lobby divides the building's flow of pathways. To the left is the office space, and to the right is the service space.

From an entry to the office, the corridor steeply curves, making one feel as if there is no end to the space. It feels like being absorbed into it. Everything in this space is white – columns, walls and even the furniture. It is quiet and tranquil that not even a breath was heard. Walking along the curved corridor and peering through the window, the vast water endlessly splashing is a pleasure to our eyes. The void space stretching up to the second floor allows for natural light to brighten the interior.

At around half way through the office space, there is a small lobby. To the left of the lobby, the meeting room protrudes towards the water. This space also has no superfluity to distract the eyes. As is the case for all indoor spaces designed by Siza, the height of ceiling in the meeting room is tiered to allow for indirect lighting to subtly brighten the interior. Some striking features of the meeting room are the table, chairs and storage furniture that Siza designed himself. The dark-brown shaded wooden furniture has an imposing presence within the completely white space.

We now turn back and head toward the main entrance to tour the opposite side of the building. The ambience of the service space is quite different from that of the office space. Unlike the bright atmosphere of the office space, the corridor in the service space exudes a somewhat dark and toned-down atmosphere with no bright lighting or windows.

A gently-sloped ramp on the right of the main entrance leads to the auditorium named "Dragon's Wings". It is a pentagon space wherein the walls that make-up the geometry are slightly concave. The ceiling looks as if a sheet has been cut out in the shape of a triangle, with one side of the triangle slightly indented. On the back wall, there is a long rectangular window placed at a height slightly lower than a person's waist. Such a design allows a person to enjoy the feeling of floating on water.

The chairman's office, secretary's office, technical areas and terrace are located on the second floor. The atmosphere is more bright and open than the first floor. Naturally, one focuses more on the scenery outside than the space inside. Factory buildings are located far from the office, thus the view opens in all directions, and nothing obstructs it.
There is a balcony on the second floor at the inner side of the U-shaped building, so that people can step outside and enjoy the view from a different angle. The gem of the second floor lies in the building's outside rather than what is found inside.

We go down to the first floor via the ramp. The ramp is quite long and its slope is gentle. There are rectangular skylights at the starting and ending part of the ramp, and midway through it, and an oval shaped window is on left and right side. Those allow people to enjoy the scene outside and experience the natural light while walking inside the ramp. This space strikes a good balance between straight and curved lines as well as between the blank ceiling and indirect lighting – something that Siza enjoys applying.

"The Building on the Water", contemplating its significance

Many people describe Álvaro Siza as the architect most capable of expressing human sensibilities through architecture. He breathes life into a simple space by maximizing the physical properties of the matter that lies within. He is also outstanding in terms of controlling the light that passes through the architecture. Siza's distinguished talent is well reflected in "The Building on the Water."
Carlos Castanheira, Álvaro Siza's partner architect, says *"There is something of the Renaissance Man in Álvaro Siza, in his wish to create sculpture, to draw and to write. This will to explore beauty revealed as an essential complement to his controlling of space and giving of form to temporal space."*
As Chairman Por-Shih Lin put it, *"Mr. Siza has always upheld the design concept of establishing harmony between man and the environment. This building ingeniously blends with the waterscape and plant grounds. As the first project in China by Álvaro Siza, it will be an inspiration for future industrial plant designs in the country."*
Shuenn-Ren Liou, professor at the Cheng Kung University, stated that "The Building on the Water" will be a turning-point for the architectural society in China. He added a meaningful comment that China needs to slow down and take a breath, as he hopes this building by Álvaro Siza will serve as momentum for the designing of other buildings that cause people to contemplate the essence of space.

Such a positive assessment of "The Building on the Water" is not limited to those mentioned above. Many people praise the work because of the significance of its "existential value" as opposed to architectural details. China was once dubbed the architectural laboratory. With the 2008 Beijing Olympics as the starting point, most parts of China's architectural society has become enthusiastic about landmark buildings that offer visual pleasure. Perhaps, the fact that many people assess Siza's project positively is an indication of hope that architects will overcome its current limitations and begin to identify with values of a different dimension.

YuMi Hyun

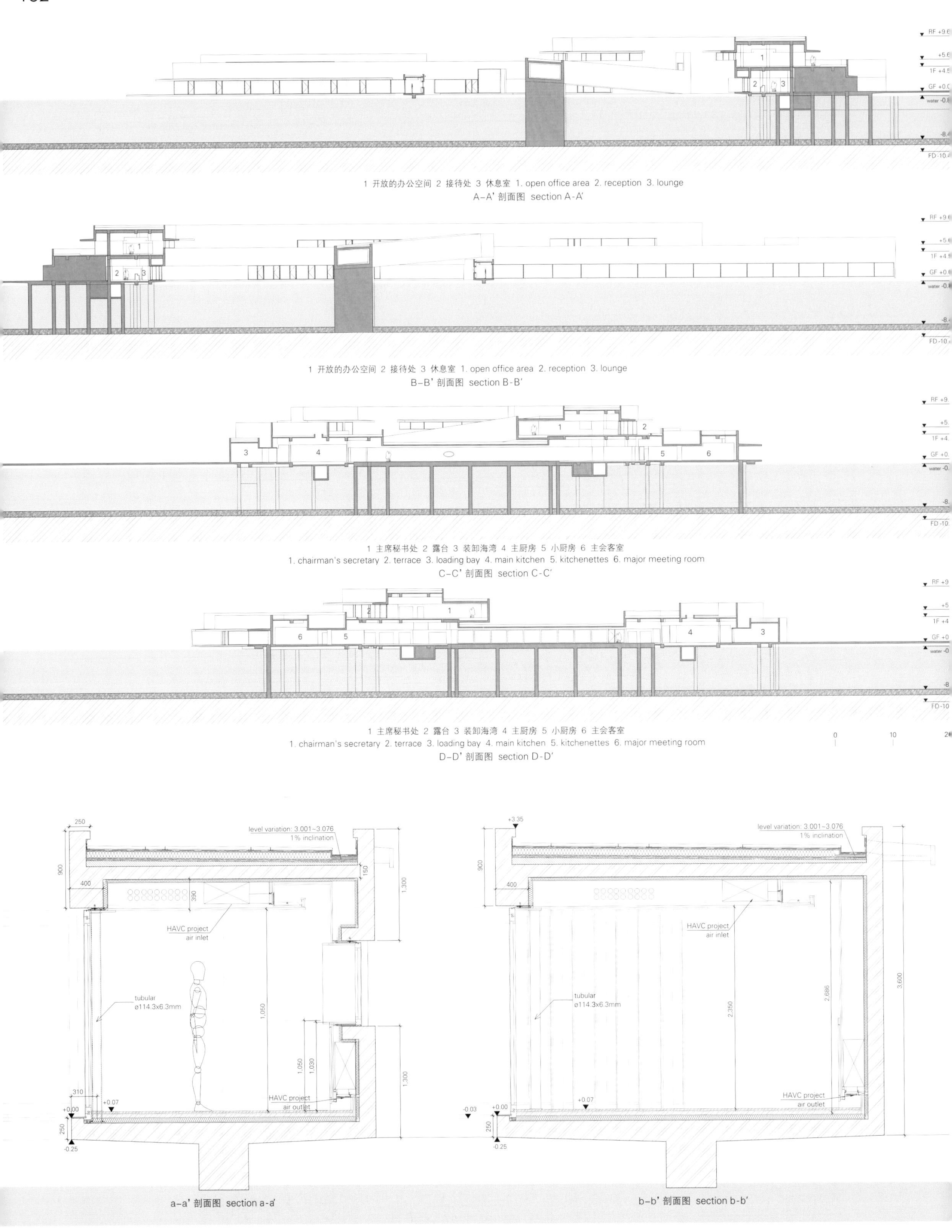

1 开放的办公空间 2 接待处 3 休息室 1. open office area 2. reception 3. lounge
A-A' 剖面图 section A-A'

1 开放的办公空间 2 接待处 3 休息室 1. open office area 2. reception 3. lounge
B-B' 剖面图 section B-B'

1 主席秘书处 2 露台 3 装卸海湾 4 主厨房 5 小厨房 6 主会客室
1. chairman's secretary 2. terrace 3. loading bay 4. main kitchen 5. kitchenettes 6. major meeting room
C-C' 剖面图 section C-C'

1 主席秘书处 2 露台 3 装卸海湾 4 主厨房 5 小厨房 6 主会客室
1. chairman's secretary 2. terrace 3. loading bay 4. main kitchen 5. kitchenettes 6. major meeting room
D-D' 剖面图 section D-D'

a-a' 剖面图 section a-a'

b-b' 剖面图 section b-b'

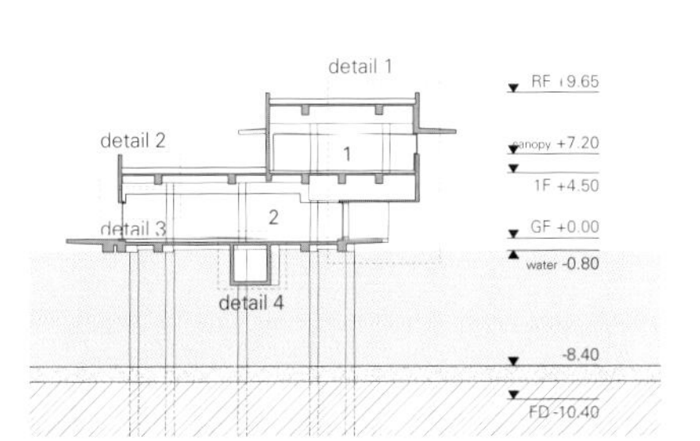

1 开放的办公空间 2 会客室 1. open office area 2. meeting room

E–E' 剖面图 section E-E'

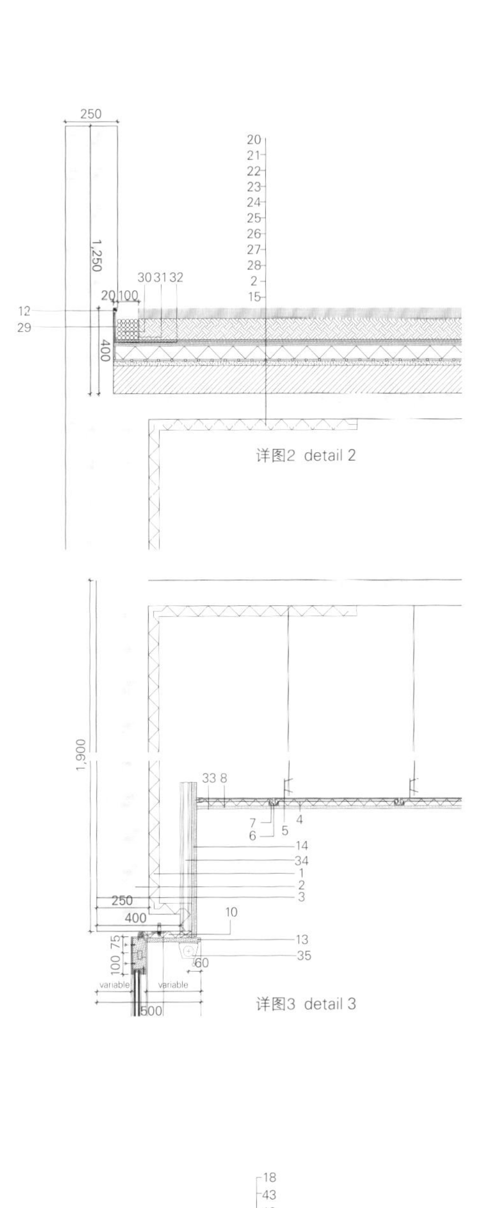

详图2 detail 2

详图3 detail 3

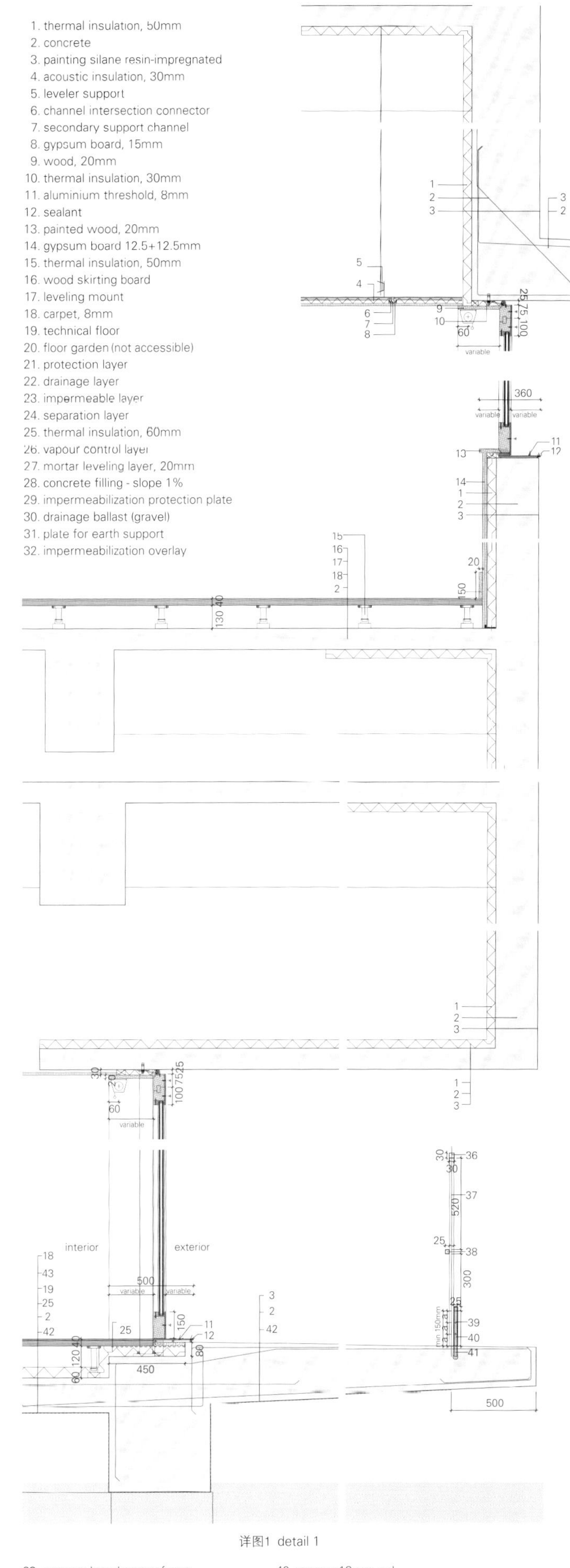

1. thermal insulation, 50mm
2. concrete
3. painting silane resin-impregnated
4. acoustic insulation, 30mm
5. leveler support
6. channel intersection connector
7. secondary support channel
8. gypsum board, 15mm
9. wood, 20mm
10. thermal insulation, 30mm
11. aluminium threshold, 8mm
12. sealant
13. painted wood, 20mm
14. gypsum board 12.5+12.5mm
15. thermal insulation, 50mm
16. wood skirting board
17. leveling mount
18. carpet, 8mm
19. technical floor
20. floor garden (not accessible)
21. protection layer
22. drainage layer
23. impermeable layer
24. separation layer
25. thermal insulation, 60mm
26. vapour control layer
27. mortar leveling layer, 20mm
28. concrete filling - slope 1%
29. impermeabilization protection plate
30. drainage ballast (gravel)
31. plate for earth support
32. impermeabilization overlay

详图1 detail 1

33. gypsum board corner frame
34. gypsum board wall studs
35. blackout
36. horizontal steel tubular 30x30x2mm
37. vertical steel tubular 25x25x2mm
38. horizontal steel tubular
39. horizontal steel tubular 25x25x2mm
40. screw ø18mm rod
41. bar welded to the rod and fixed to concrete with chemical anchor
42. superfine epoxy cement sealing mortar and temporary moisture barrier, 2mm
43. plywood
44. concrete screed 60mm

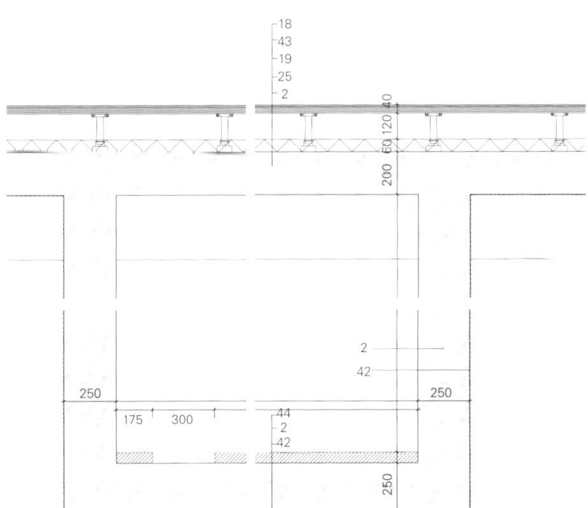

详图4 detail 1

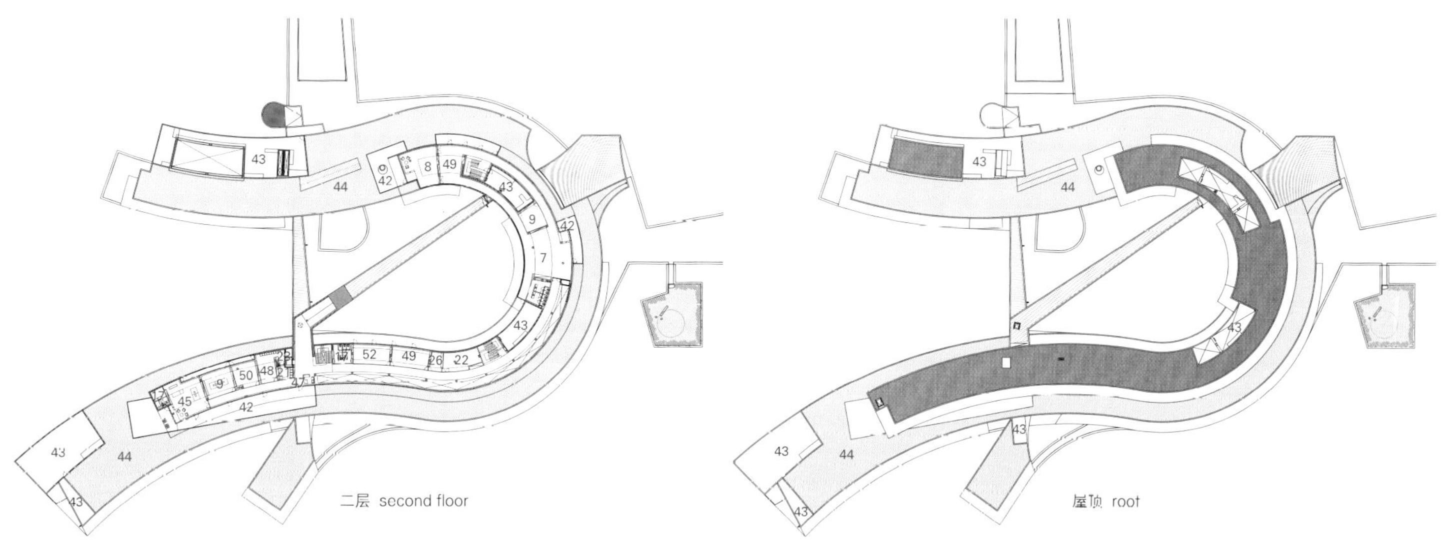

二层 second floor

屋顶 roof

1 主入口
2 接待处
3 休息室
4 会议室
5 VIP会客室
6 展览室
7 开放的办公空间
8 私人办公室
9 会客室
10 主会客室
11 董事长办公室
12 财务总监办公室
13 带有保险箱的财务档案室
14 人力资源文件储存室
15 销售文件储存室
16 采购文件储存室
17 卫生间
18 电气供应室
19 应急供应室
20 安保控制室
21 酒和礼品储存室
22 机房
23 无线电通讯室
24 小厨房
25 通用储存室
26 会客室/客人等候室
27 服务入口
28 主厨房
29 餐饮服务室
30 洗碗室
31 储存室
32 面粉房
33 员工更衣室
34 餐馆的开放空间
35 私人用餐室
36 VIP用餐室
37 装卸海湾
38 办公室
39 员工休息室
40 垃圾室
41 阳台/室外流线
42 露台
43 技术区
44 绿化区
45 主席办公室
46 CEO办公室
47 休息室
48 主席秘书处
49 秘书处
50 包间

1. main entrance
2. reception
3. lounge
4. conference room
5. VIP meeting room
6. show room
7. open office area
8. private office room
9. meeting room
10. major meeting room
11. director's office
12. finance director's office
13. finance archive with safe box
14. HR files storage
15. sales files storage
16. purchasing files storage
17. toilet
18. electric supply room
19. emergency supply room
20. security monitoring room
21. wine and gifts storage
22. computer room
23. telecommunications room
24. kitchenettes
25. general storage
26. meeting room / guests waiting area
27. service entrance
28. main kitchen
29. food serving
30. dish washing
31. storage
32. flour workshop
33. employees locker room
34. restaurant's open space
35. private dining room
36. VIP dining room
37. loading bay
38. office
39. employees lounge
40. trash
41. balconies/exterior circ.
42. terrace
43. technical areas
44. green
45. chairman's office
46. CEO office
47. lounge
48. chairman's secretary
49. secretary
50. private room

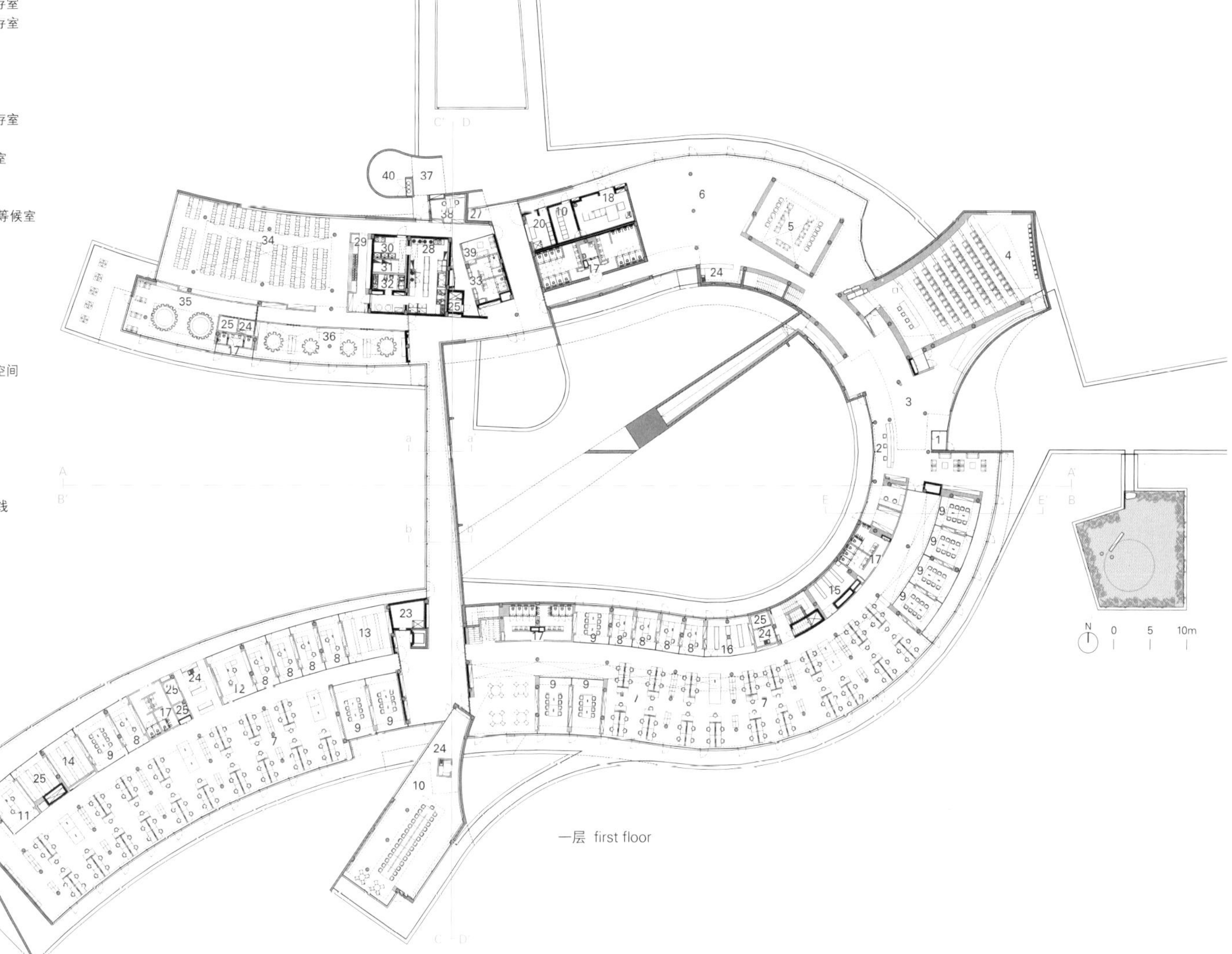

一层 first floor

项目名称：The Building on the Water
地点：Huai'an City, Jiangsu Province, China
建筑师：Álvaro Siza, Carlos Castanheira
主要负责人：Pedro Carvalho, Luís Reis(first phase), Luís Reis(second phase)
合作者：Diana Vasconcelos, Susana Oliveira, Elisabete Queirós, Orlando Sousa, Rita Ferreira, João Figueiredo, Caitriona, Anand Sonecha
3D模型和绘图：Francesco Sechi, João Figueiredo, Pedro Afonso, José Soares
创意集成顾问：Xue Xue Institute, Xue Xue Foundation
项目经理/施工监理：
Stephen Wang & Richard Wang with Chiou-Hui Lin
本地建筑师和工程师：
United Architects & Engineers Co., Ltd
顾问：structure_HDP–Paulo Fidalgo/ mechanical_GET–Raul Bessa/ lighting_GPIC–Alexandre Martins
施工单位：Zhejiang Urban Construction Group
甲方：Shihlien Chemical Industrial Jiangsu Co., Ltd
总建筑面积：8,600m² / 有效楼层面积：11,000m²
设计时间：2010 / 施工时间：2011 / 竣工时间：2011
摄影师：©FG+SG Architectural Photography

持续且不断发展的传统

Persistent and Ev

在最近几年，全球化现象的增长和普遍的城市扩张导致人们越来越关注挽回当地文化的现象，而这些文化在某种程度上能够定义一种身份。这不能被看作是一种排外的条件，而是一种多重性条件，也是一种丰富的体验。

在建筑方面，这种寻求被遗忘的身份的需求就意味着要回望乡土建筑，并了解它如何可以促进当今每个社会的现代化需求。

在规模较小的项目中，这个过程肯定更容易，如在私人住宅设计中，但是大型投资企业集团就要都出局了。在经济利益和市场策略下，这确保了个人的利益和感情。

所展示的项目全部位于亚洲，中东、中亚、东亚和东南亚地区。它们是应用不同方法和结果的实例，这种独特的地方传统可引入当代建筑项目中。那些无数的变数包括从每一种文化发展的差异化生活体验、历经对特定气候特点的应付方式、当地材料的使用和施工方法，到处理文化和历史遗产的方式。

The growth of the globalization phenomenon and the generalized urban expansion resulted, during the recent years, in an increasing concern about retrieving the local cultural aspects that can somehow define an Identity. This could not be seen as a condition of exclusion but, on the contrary, of plurality and as an enriching experience.
In architecture this quest for that forgotten Identity can mean to look back to the vernacular architecture and learn how it can contribute today to the contemporary needs of each society.
This process is surely easier in smaller scale projects, like in the private dwelling design, where the big investment group corporations are out of the game. This guarantees that the personal interests and affections are taken into account over economic interests or market strategies.
The presented projects are all located in Asia, from the Middle East, Central, East and Southeast Asia. They are examples of the diverse approaches and results that such distinct local traditions can be induced to the contemporary architecture project. Those numerous variables run from the differentiated life experiences developed by each culture, going through the answers to specific climate characteristics, the use of local materials and construction methods, to the way of dealing with cultural and historical legacies.

微胡同_Micro-Hutong/standardarchitecture
雅法厂房_Factory Jaffa House/Pitsou Kedem Architects
石材浇筑的住宅_The House Cast in Liquid Stone/SPASM Design Architects
古梅斯苏别墅_Gumus Su Villas/Cirakoglu Architects

持续且不断发展的传统_Persistent and Evolving Traditions/Paula Melâneo

olving Traditions

我们可以带着瑞士民族学家和建筑人类学家Nold Egenter的一些问题来介绍这一主题。在他的《民居建筑——象征意义从何而来》中，有一些关于住宅的人类学注释。

"未来的人类准备好居住在宇宙内的同种性质的物理空间了吗？我们对现代建筑师和城市规划专家从天文学和物理学借来的空间概念感到满意吗？还是我们应该再次尝试理解人类的遗产、乡土建筑极致且和谐的表达？

我们是否满意建筑师投射到我们的城市和乡村的摩天大楼和天际线中的、合理且统一的几何类型？还是我们想要重新发现和谐的建筑形式的传统方式成为平衡人类生活的典范？如果我们发现建筑在很大程度上创造了人与文化，可能有合理的动机比今天更严肃地研究乡土建筑。"

现代化的建筑和都市主义提供的统一的生活理念和合理的空间通过时间证明是不可靠的。在近期实行的普遍标准的城市化生活方式并不适合所有的一切。

这导致了我们需要及时回头，用传统和乡土的方法寻求几乎被建筑与施工遗忘，甚至失去了的根。然而，这并不意味着乡土化结果，而是研究和学习的过程，整理重要的传统解决方案，这一方案需要等待成熟的时间，并证明是有效的，因此这不应该是微不足道的，相反，应该是要保留下来的。

时下，这种情况贯穿到一些社会活动中，在这些活动中，文化和商业化生产的低俗性减少了地方特色。这就是为什么我们目前正面临着古老传统的复苏，如生产方法、当地材料和传统品牌的使用。

建筑抵消了一种趋势，即发展无菌的、通用的建筑概念或构建具有形式主义特征的壮观建筑，它们尊重这些假设的、往往是充满了信息和参考的趋势，但同时更谦卑。它们更深刻地涉及了自然和现有的本地建筑，并且整合可能指向本地身份的奇异元素。

整个过程和结果介于民居建筑遗产之间，是"追求地域特色"的理

We can introduce the theme with some of the questions Nold Egenter, Swiss ethnologist and architectural anthropologist, poses in his "Vernacular Architecture – Where do the symbolic meanings come from?" Some notes are regarding the "anthropology of the house":

"Are human beings of the future prepared to live in the homogeneous space of physics, of the universe? Are we pleased with the space concepts modern architects and urbanists borrowed from astronomy and physics? Or should we try again to understand the human heritage, the categorically polar and harmonious expression of vernacular architecture?

Are we happy with the uniform geometry type of skyscrapers and rationalistic skylines architects project into our cities and villages? Or do we want to rediscover the traditional way of harmonising architectural forms as a model of balanced human lives? If we discover that architecture to a great extent created man and culture, there might be reasonable motives for studying vernacular architecture more seriously than this is done today."

The homogeneous life concepts and rational spaces that modern architecture and urbanism have offered proved through time to be fallible. Also, the universal standardized way of living that urbanity has imposed in the recent times is not suitable for all.

This has led to go back in time and seek for the almost forgotten or even lost roots of architecture and construction, in its traditional and vernacular approaches. Yet, it doesn't mean a vernacular outcome, but a process of studying and learning, sorting important traditional solutions that time matured and proved to be effective, and which therefore should not be negligible but, instead, preserved.

This situation is nowadays transversal to several activities in society, where the vulgarization of cultural or commercial production muted the local characteristics. That's why we are currently facing the recovery of ancient traditions, such as production methods, the use of local materials or traditional brands.

Counteracting a tendency to develop aseptic universal architectonic concepts or construct spectacular architecture of formalist character, the architecture that respects these assumptions tends to be full of information and references, but is more humble at the same time. It is more deeply related with nature and with existing local architecture, integrating singular elements that could point a local Identity.

The whole process and result lie somewhere between the ver-

照片提供：©SPASM Design Architects

石材浇筑的住宅，印度Khopoli
House Cast in Liquid Stone, Khopoli, India

雅法厂房，以色列特拉维夫
Factory Jaffa House, Tel Aviv, Israel

照片提供：©Pitsou Kedem Architects(Amit Geron)

念”，追随了诺伯格–舒尔茨的现象学方法和肯尼思·弗兰普顿的批判地方主义。没有体现出持续性的本地化和传统正在恢复，但作为一个重新的诠释起到了重要的意义。当地的环境，如地形、气候、大气、经济和社会文化特色，其他还有很多，都要考虑入内，以重新恢复身份和与地方之间更强烈的关系。这就产生了建筑更熟悉且更有情感的方式，其中之一便是更接近个体。

在本章所选择的项目中，传统、时尚的学习方式和生活经历就是建筑师采用特殊的方式谱写新的历史所使用的词典和词汇，来协调地方特色、甲方的意愿和建筑语言本身。

最近完成的、由Cirakoglu建筑师事务所建造的古梅斯苏别墅位于Gümüşlük村，这是土耳其地中海西海岸的一个平静且安静的港湾。周围景观的优势、石灰岩覆盖着地中海的山地形象，已经激发了该地区的传统建筑。伊斯坦布尔地区的办公室也采用石头作为建筑材料，这不仅考虑到当地的美感，而且考虑到能源效率。塑造别墅外形的厚墙壁提供了热质量和保温性能，同时规划了住宅的每个房间。这一系列的小单元由开放式的走廊穿过，空气在此循环，用作自然冷却和通风系统。大型滑动玻璃表面划分空间，并把周围美丽的全景囊括进来。石材、木材和玻璃都是主要使用的材料。使用干燥的藤茎屋顶（这里选择的是竹藤条）是地中海的传统，保证室外天篷提供遮阴，同时保护暴露在强烈的阳光下的住宅。

石材浇筑的住宅是对岩石景观环境的现代解读，它位于印度马哈拉施特拉的西部高止山脉（高地）。通过在施工中使用清水混凝土这个“现代”的材料来代替石头，SPASM设计建筑师事务所想要他们的项目与周围的黑色玄武岩环境融合：“当地黑色玄武岩就是这个地方的特色。”住宅被认为是一个庇护所，以抵挡强烈的气候变化，如猛烈的季雨和夏季的炎热，它作为一个安全的庇护所，拥有强大的存在感，是一处宏伟和鼓舞人心的空间，来让人们欣赏风景。

Pitsou Kedem建筑师事务所设计的雅法厂房坐落在地中海的一个

nacular architecture legacy, the idea of Genius Loci, following Norberg-Schulz's phenomenological approach and Kenneth Frampton's Critical Regionalism. The vernacular and traditions are recovered without reflecting a continuity, but as a reinterpretation taking a new significance. Local context such as topography, climate, atmosphere, economic and socio-cultural characteristics, among many others, are taken into account to retrieve an Identity and a stronger relation with the place. This results in a more familiar and affective approach on architecture, one that is closer to the individual.

In the selected projects for this chapter, the tradition, popular learning and life experiences work as a lexicon or vocabulary that is used by architects in specific ways to write a new history, conciliating the local characteristics, the clients' will and the architecture language itself.

Recently finished, Cirakoglu Architects' project for the Gumus Su Villas is located in Gümüşlük, a calm and quiet harbour in Turkey's Mediterranean west coast. The strength of the surrounding landscape, the image of the hills over the Mediterranean Sea covered in limestone, already inspired traditional architecture in the area. The Istanbul based office also adopted the stone as a construction material considering not only the local aesthetics but also the energy efficiency. The thick walls that shape the villas provide the thermal mass and insulation, while outlining each room of the house. This series of small units is crossed by open corridors, where the air circulates, functioning as a natural cooling and ventilation system. Large sliding glazed surfaces divide the spaces and bring inside the beautiful surrounding panorama. Stone, wood and glass are the main materials used. The dried cane roofing (here the option was for the bamboo cane), is a Mediterranean tradition and guarantees the shade under the exterior canopy, protecting the dwelling from the strong sunlight.

The House Cast in Liquid Stone is a contemporary reinterpretation of a rocky landscape context, located in the Indian Western Ghats (highlands), Khopoli, in Maharashtra. By using exposed concrete, the "modern" material that replaced stone in construction, SPASM Design Architects wanted to merge their project with the dark basaltic surrounding: "The local black basalt rock of the region is what this site was about." The house is thought to be like a shelter to meet the strong climatic changes, between the violent rain of the monsoon and the heavy heat of the summer. It acts as a safe refuge with a strong presence, a great and inspiring space to en-

微胡同，中国北京
Micro-Hutong, Beijing, China

古老的港口，这个港口今天已经合并到以色列的特拉维夫市区，这个项目是一个富有表现力的、拥有几百年历史的石灰石建筑街区的嵌入结构，跨越宗教、政治、文化和商业活动的范围。建筑师通过拆除其原有状态上的所有特色，来恢复建筑。这种"过去的苦行僧模式"是新结构的基础。在维护主体结构方面，正如他们所指："保护建筑物外壳的纹理和材料，并遵守建筑物工程协议"，新旧之间的差异故意设计得招摇。现代材料，如铁、不锈钢、可丽耐材料和混凝土装饰了内部，实现空间流动的新颖性和和谐性，可以很容易地应付目前生活方式的日常需求。

微胡同是总部位于北京的标准营造建筑事务所的一个实验。胡同是典型的老北京高密度街区类型。这一实践采取了批评的态度来对待这种传统的住宅空间。其实，它们今天已经过时并被当地居民抛弃，因为它们无法满足人们的当代需求，在现有的结构规模相同的嵌入结构内，微胡同是一个小社会，占据了大栅栏地区胡同里的一个内部庭院，新建的木体量动态地交织在一起。这个项目的意义在于深入研究传统空间特色和本地文化的知识，产生了一个重新思考原来的占地计划的重要方法，以对当代的要求做出回应。

作为本章的特色项目，在任何审美潮流中，不那么正式的质量使其寿命更加长久。因此，这种建筑可以被改造。在大部分的时间内，它成为一个更可持续的模式，以延长寿命，或成为一个传统，且地方特色和解决方案对于特殊的地方要求来说，是一个稳定的答案。

joy the scenery.

Factory Jaffa House by Pitsou Kedem Architects, is located in an ancient port over the Mediterranean Sea, today integrated in Tel Aviv's city area, in Israel. This project is an intervention over a centuries-old limestone building of a neighbourhood loaded with an expressive historic legacy, crossing religion, politics, cultures and commercial activity. The architects rehabilitated the building by demolishing all features made over its original status. This "ascetic style of the past" was the base for the new intervention. Maintaining the main structure, as they refer: "preserving the textures and materials of the building's outer shell and respecting the building engineering accord", the differences are intentionally notorious between the new and the original. Contemporary materials such as iron, stainless steel, corian and concrete, equip the interior achieving a new harmony of the spaces' flow that can easily answer to daily needs of present life style.

Micro-Hutong is an experiment by Beijing based practice standardarchitecture. Hutongs are a typical and dense typology of old Beijing's neighbourhoods. This exercise treats the tradition by taking a critical approach towards this kind of traditional housing spaces. In fact they are becoming obsolete structures today and abandoned by original dwellers, as they are unable to meet their contemporary demands. Within the same intervention scale of the existing constructions, the Micro-Hutong is a small social space occupying an interior courtyard of this Dashilar District's hutong, with a dynamic game of new wooden volumes. This project's relevance lies in the deep study and knowledge of the traditional space's characteristics and local culture, with a critical approach capable of rethinking the original occupation program in order to respond to the contemporary requirements.

As in the projects featured in this chapter, the quality of being less formal or within any aesthetic trend makes it more perennial. Thus, this kind of architecture can be translated, most of the times, as a more sustainable model, with an enlarged lifespan, or as a tradition, local features and solutions are a stable answer to most specific local needs. Paula Meláneo

微胡同

standardarchitecture

微胡同是张柯的标准营造建筑团队在大栅栏地区的杨梅竹街道进行的一个建筑实验。该项目的目标是在北京密度超大的传统胡同空间的限制内，寻找创造超小规模的社会住房的可能性。

坐落在大栅栏地区，步行即可抵达天安门广场内的历史区，标准营造建筑事务所设计了一个30m²的微型胡同，成为一个全新的保护和建造胡同的替代品。

对胡同动态的严格审视表明，即使有不法房地产开发商来势汹汹地占地，胡同的最关键问题还是其居住者的无情离去。考虑到设施的缺乏和优质的公共空间的缺失，他们决定将其出售并搬到市郊更大的公寓中。这种在北京心脏的传统胡同里居住的居民不断遗弃胡同的现象加速了一种策略的形成，以挑战租户对胡同的日益冷淡，使脆弱的生存条件变得生机勃勃。

其结果是建造一座建筑，把作为生产功能机器的庭院重新带回来，因为它通过创造与城市环境之间的直接联系，来激活建筑，并牵引到它的内部社交活动中。除了加强通风和照明，庭院创造了动态体量内的居住空间和建筑前方的门廊之间的直接关系。这种灵活的城市生活房间成为私人房间到街上的过渡区，同时也成为微胡同的居民和社区邻居使用的半公共空间。

微胡同继承了传统胡同的比较密的规模，重新激活社会的冷凝氛围，同时通过提高空间质量来增强空间感。它的轻型钢结构和胶合板面板覆层允许低成本建设，同时为北京胡同的未来创造全新的、可能的重新布局。

Micro-Hutong

Micro-Hutong is a building experiment by Zhang Ke's Standardarchitecture team on the Yangmeizhu Street of Dashilar area. The goal of the project is to search for possibilities of creating ultra-small scale social housing within the limitations of super-tight traditional hutong spaces of Beijing.

Located in the Dashilar District, a historical area within walking distance of Tiananmen Square, standardarchitecture has designed a 30-square meter Micro-Hutong that offers a new alternative to Hutong preservation and actualization.

A critical look into the dynamics of the hutong reveals that even with the menacing grip of unscrupulous real estate development, the most critical problem of the hutong consists of the relentless exodus of its occupants. Concerned with the lack of facilities and the absence of quality communal space, they decide to sell and move out to bigger apartments outside of the city center. This

照片提供: courtesy of the architect

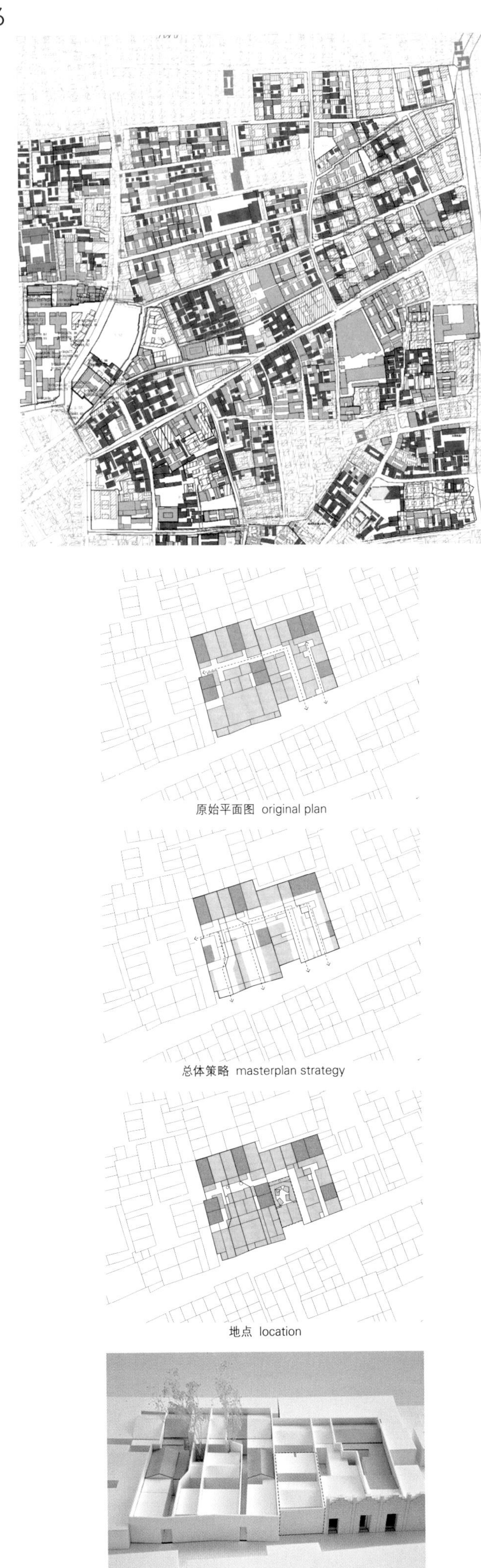

原始平面图 original plan

总体策略 masterplan strategy

地点 location

总体模型 masterplan model

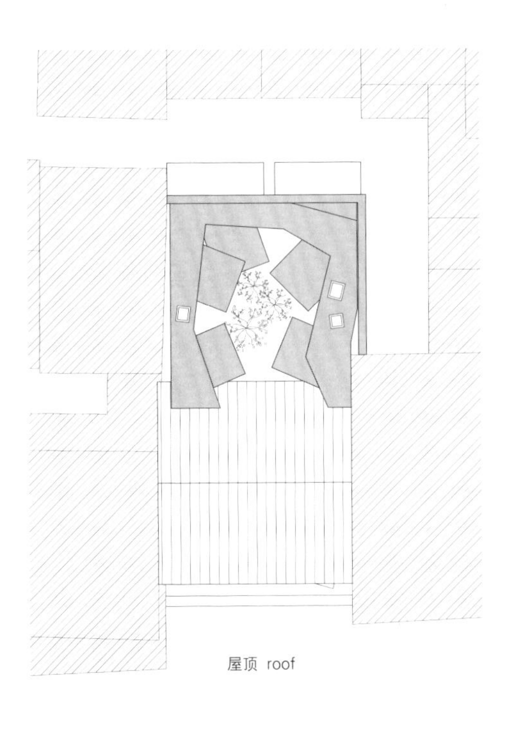

屋顶 roof

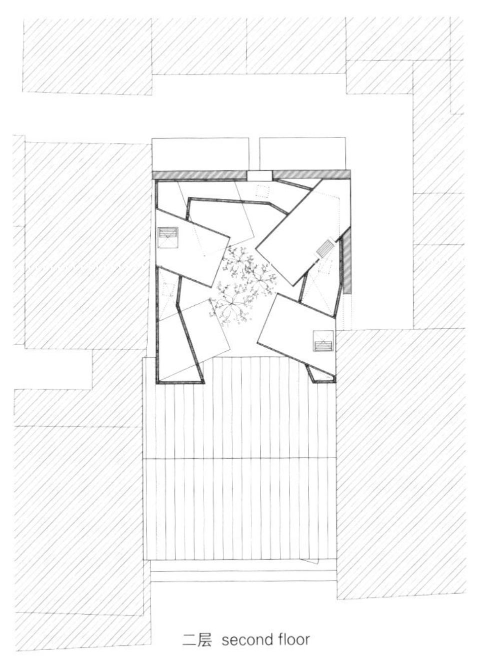

二层 second floor

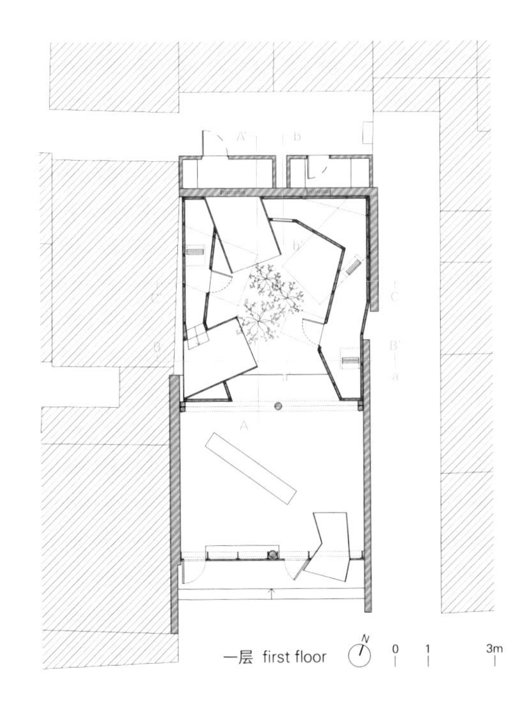

一层 first floor

光纤宽带
已覆盖
1—15

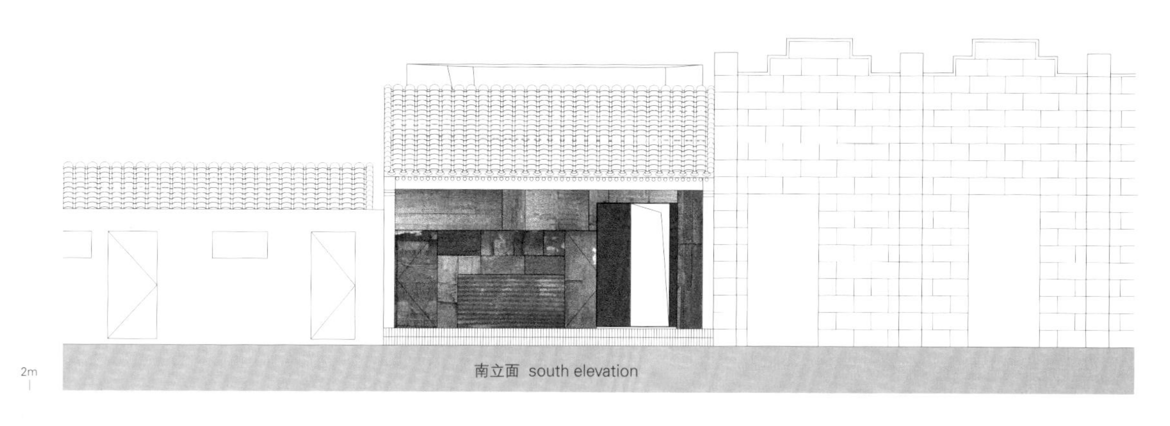

南立面 south elevation

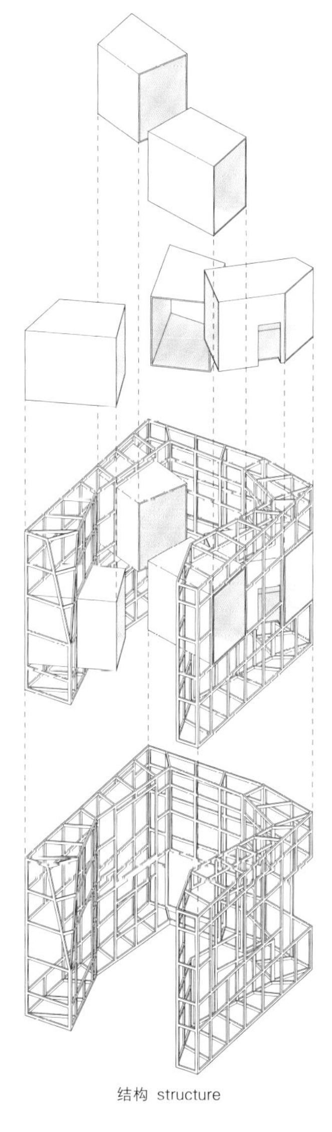

结构 structure

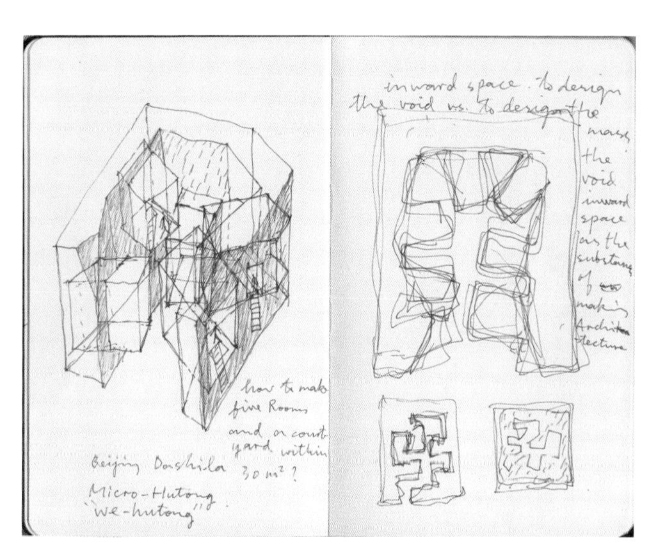

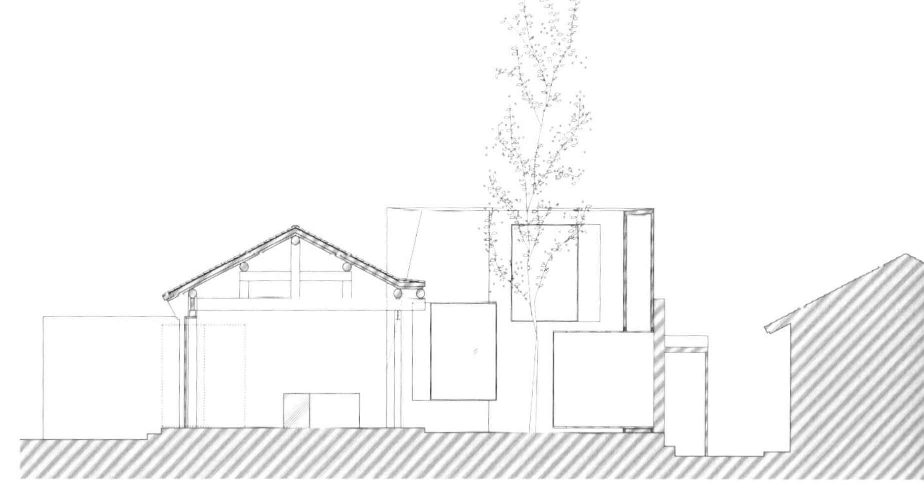

A-A' 剖面图 section A-A'

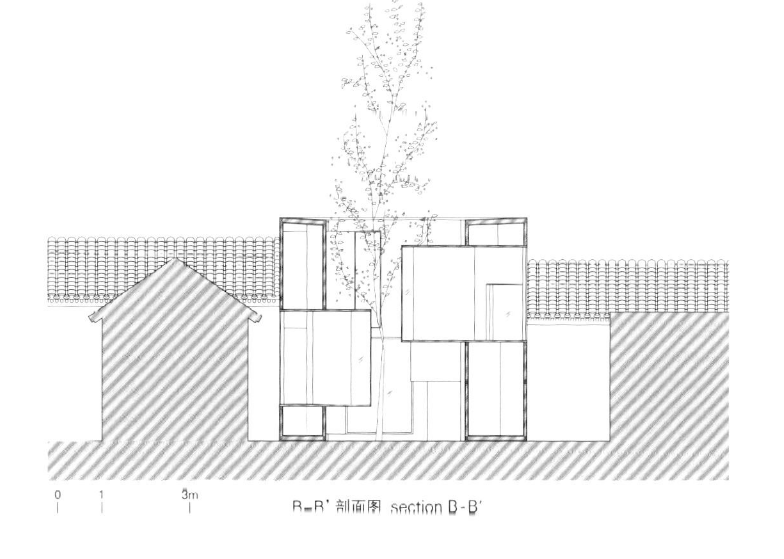

B-B' 剖面图 section B-B'

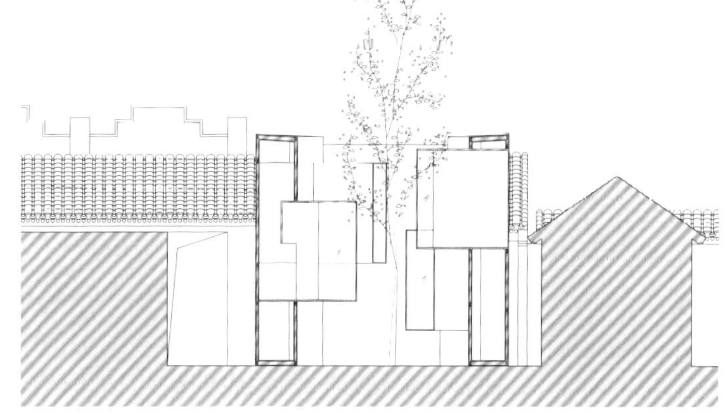

C-C' 剖面图 section C-C'

constant desertion of the traditional dweller of the hutong from the heart of Beijing prompted to generate a strategy able to challenge the growing disinterest of the hutong tenants in order to keep alive valuable living traditions.

The result is an architectural operation that brings back the courtyard as a generator of program, as it activates the building by creating a direct relationship with its urban context, drawing to its interior social activities. Apart from enhancing the flow of air and light, the courtyard creates a direct relationship between the living space contained in the dynamic volumes and an urban vestibule in the front part of the building. This flexible urban living room acts as a transition zone from the private rooms to the street, while serving as a semi-public space to be used by both the inhabitants of the Micro-Hutong and the neighbors of the community.

The Micro-Hutong inherits the intimate scale of the traditional hutong, revitalizing its social condensing capabilities, while enhancing it with spatial improvements. Its light-steel structure and plywood panel cladding allow for low-cost construction, while creating new possible reconfigurations for the future of the Hutong in Beijing.

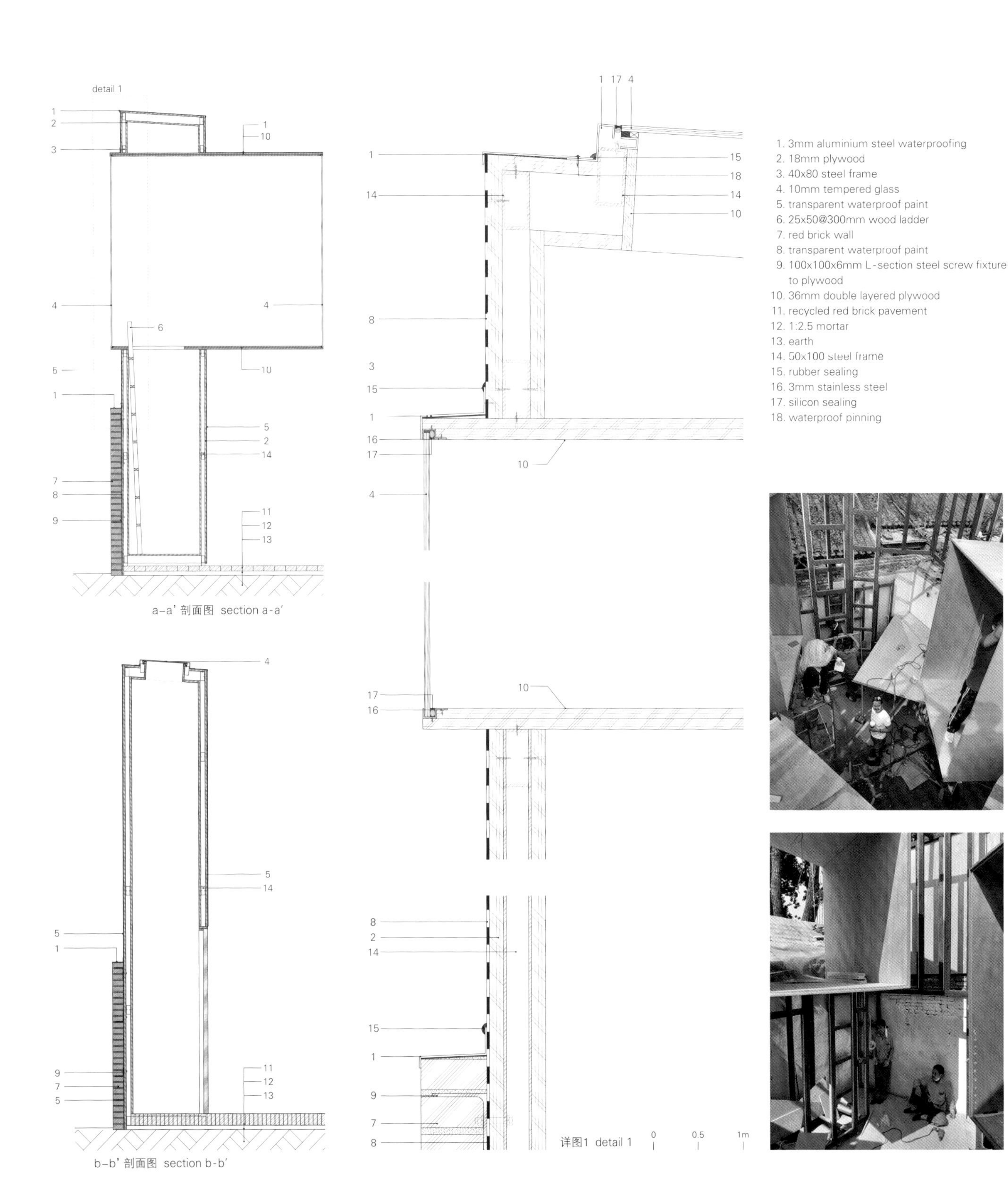

a-a' 剖面图 section a-a'

b-b' 剖面图 section b-b'

详图1 detail 1

项目名称：Micro-Hutong
地点：Yangmeizhu Street, Dashilar, Beijing, China
建筑师：standardarchitecture
项目建筑师：Zhang Ke, Zhang Mingming
项目团队：Dai Haifei, Ao Ikegami, Zhang Yanping, Huang Tanyu, Zhang Zhaosong
用地面积：30m² / 总建筑面积：48m² / 有效楼层面积：60m²
设计时间：2012.3—2013.9 / 施工时间：2013.9
竣工时间：2013
摄影师：
©Zhang Mingming(courtesy of the architect) - p.148, p.155 left-middle, right-top, right-bottom
©Chen Su(courtesy of the architect) - p.145 bottom, p.147, p.149, p.150, p.153, p.154, p.155 left-top, left-bottom

雅法厂房

Pitsou Kedem Architects

嵌入老雅法历史住宅中的极简主义语言

该180m²的住宅位于老雅法。它的位置是独一无二的，因为它位于海港之上，朝西，所有的洞口都面朝地中海壮丽的景色。虽然难以确定建筑物的确切年龄，但是很显然的是，它拥有几百年的历史。多年来，它已经发生了许多变化，且增建了许多结构，这些结构损害了建筑物及其空间的原有品质。项目的中心理念是恢复建筑的原始面貌、特点、石墙、分割的天花板和拱门，包括原始材料（陶器和海滩沙子的组合）。该建筑的较新的墙体覆层已被清除所有多余的元素，且经历了剥离的过程，以暴露其原始状态。出人意料的是，现代且简约的建筑风格使我们想起了过去的苦行僧模式，且与之相对应，尽管它们之间有着巨大的时间差。项目的中心思想是新旧结合，同时保持每部分的质量，并创造新的空间，将所有的风格混合甚至进行突出，因为它们具有不同时期之间的对比度和张力。历史通过保存建筑物外壳的纹理和材料，并且遵守建筑物工程协议来表现出来。其现代性通过空间的洞口、改变内部流动形成一个更为开放且自由的环境以及在各分区、各洞口和各家具中应用不锈钢、铁来表达。该项目成功地尊重和保护结构的历史以及浪漫价值，同时建造了现代工程，具有现代性且适用于这个时期。尽管时代不同了，但不同时期的张力和对立依然在一处令人惊讶的平衡且和谐的空间内存在着。

Factory Jaffa House

The Language of Minimalism Embedded in a Historic Residence in Old Jaffa

The 180-square-meter residential home is located in Old Jaffa. Its location is unique in that it is set above the harbor, facing west with all of its openings facing the majestic splendor of the Mediterranean Sea. Whilst it is difficult to determine the building's exact age, it is clear that it is hundreds of years old. Over the years, it has undergone many changes and had made many additions that have damaged the original quality of the building and its spaces. The central idea was to restore the structure's original, characteristics, the stone walls, the segmented ceilings and the arches including the exposure of the original materials (a combination of pottery and beach sand). The building has been cleaned of all of the extraneous elements, from newer wall coverings and has undergone a peeling process to expose its original state. Surprisingly, modern, minimalistic construction styles remind us of and correspond with the ascetic style of the past, despite the vast time difference between them. The central idea was to combine the old and the new whilst maintaining the qualities of each and to create new spaces that blend the styles together even intensify them because of the contrast and tension between the different periods. The historical nature is expressed by preserving the textures and materials of the building's outer shell and by respecting

NETIV HAMAZALOT
לנמל
TO THE PORT

the building engineering accord. The modernity is expressed by the opening of spaces and by altering the internal flow to one more open and free environment along with the use of stainless steel, iron in the various partitions, in the openings and in the furniture. The project succeeds in both honoring and preserving the historical and almost romantic values of the structure whilst creating a contemporary project, modern and suited to its period. Despite the time differences, the tensions and the dichotomy between the periods exist in a surprisingly balanced and harmonic space.

项目名称：Factory Jaffa house
地点：Tel Aviv, Israel
建筑师：Pitsou Kedem Architects
项目团队：Pitsou Kedem, Raz Melamed, Irene Goldberg
用地面积：180m²
设计时间：2012
施工时间：2013
竣工时间：2013
摄影师：©Amit Geron (courtesy of the architect)

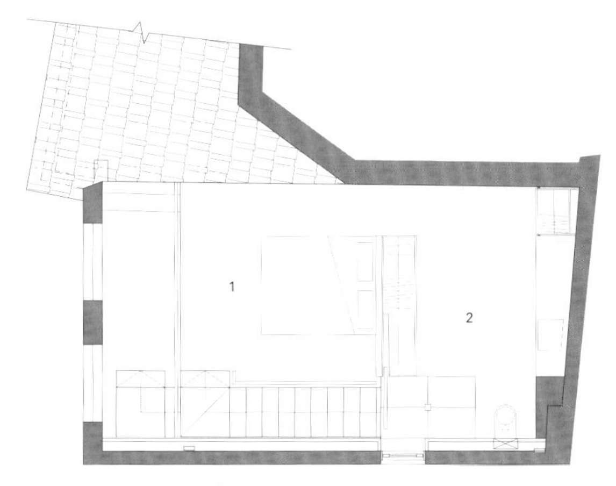

1 卧室 2 卫生间 1. bedroom 2. w.c.
三层 third floor

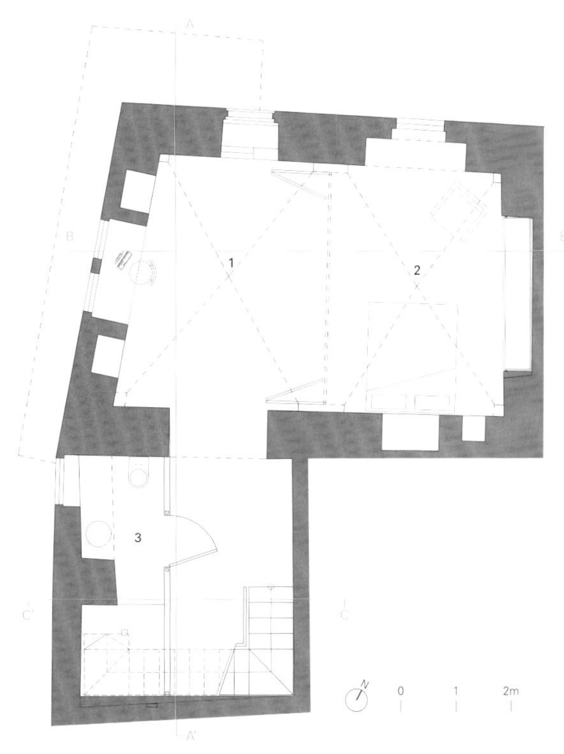

1 入口大厅 2 客房 3 卫生间 1. entrance hall 2. guest room 3. w.c.
一层 first floor

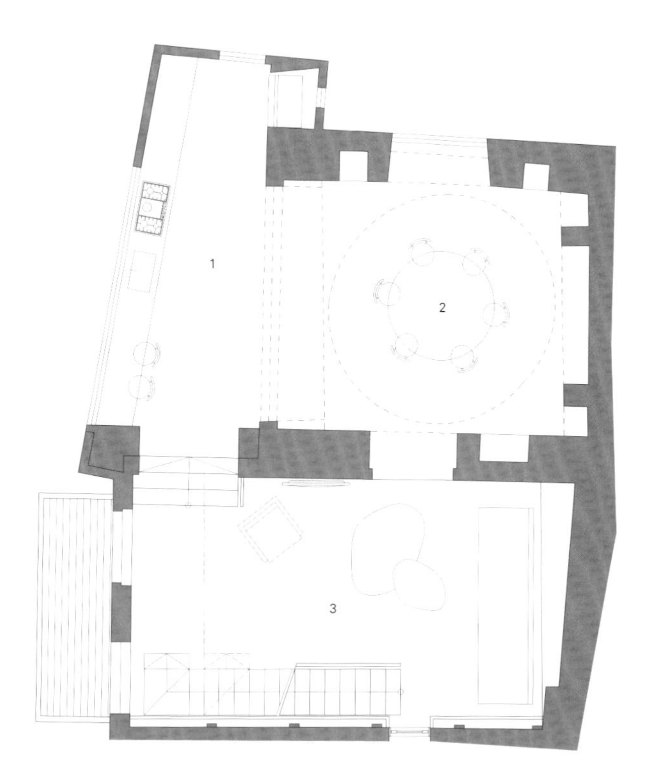

1 厨房 2 餐厅 3 起居室 1. kitchen 2. dining room 3. living room
二层 second floor

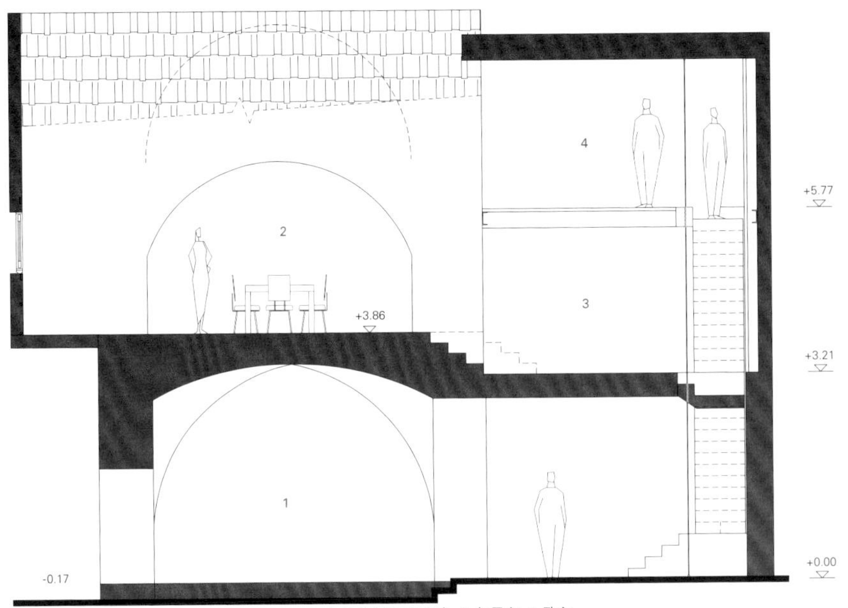

1 入口大厅 2 厨房 3 起居室 4 卧室
1. entrance hall 2. kitchen 3. living room 4. bedroom
A-A' 剖面图 section A-A'

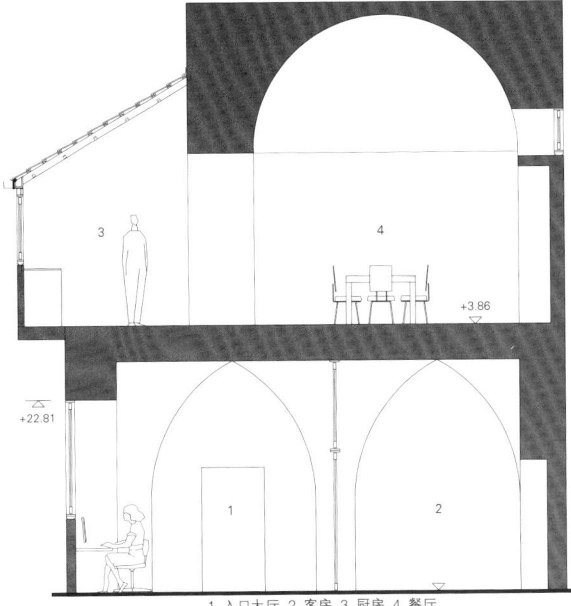

1 入口大厅 2 客房 3 厨房 4 餐厅
1. entrance hall 2. guest room 3. kitchen 4. dining room
B-B' 剖面图 section B-B'

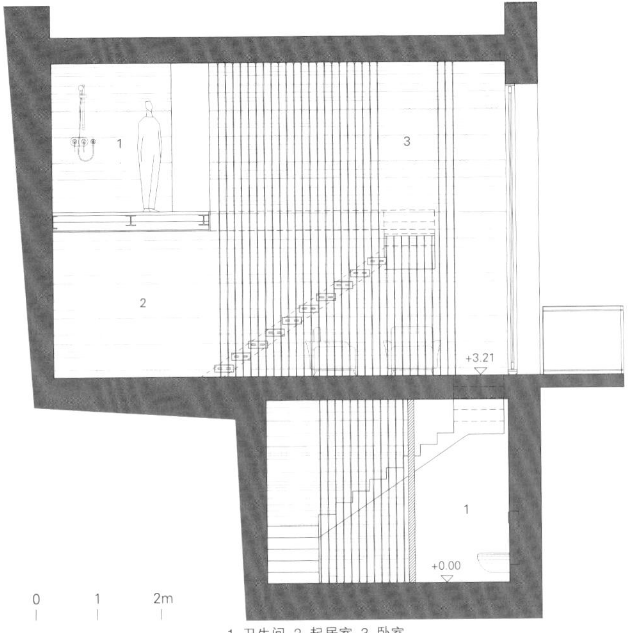

1 卫生间 2 起居室 3 卧室
1. w.c. 2. living room 3. bedroom
C-C' 剖面图 section C-C'

石材浇筑的住宅

SPASM Design Architects

又一座住宅坐落于印度马哈拉施特拉khopoli镇的西部高止山脉的起点上，这是一处到处都是岩石的地面。在季风时期，这里的降水量很大，而夏季受热均衡，该场地从3月到7月随着西南季风的爆发变化很大。

黑色玄武岩便是这一场地的地貌。建筑师选择在这片露出地面的玄武岩层上建造住宅，并将其作为一个冲积层。它们都自然地进化成水、沙子、水泥和颗粒状玄武岩的混合状态。混凝土作为避难所的材料，经历了很好的磨练，能够面对这一地区的气候变化。

这座住宅被设想为人类所占据的铸件，也是一处将景色、太阳、雨、空气混合在一起的场所，位于悬崖边上。屋顶和墙壁类似于生长的珊瑚，决定居住于此所带来的经验。

基于石材的使用、磨损、固定方式以及纹理等性质，石头可用于多种形式。饱和的无光泽黑色特质创造出一种庇护和平静的感觉。

照片无法表现出重量感，也无法表现出露天露台远端的泳池边的释放感，更深刻的途中探索感要穿过堆积的石头，直入低层的卧室中。

这所住宅可以被看做是一个铸件，是一个可以居住的物体，不管你怎么称呼它，都能被转化成一个瞭望台，让人时时刻刻地观察和感受大自然的奇观和绿荫的美景，并将它们作为抵抗强烈的热带阳光的休闲之所，同时也使生活恢复活力，"爆发"出一片绿荫。当强硬的、带有冲击性的季风到来时，微风飘荡，空气中充满湿润的泥土的芬芳，星星也在非常黑暗的夜空中移动。

建筑师没有建造一座戏剧性的建筑，却要通过我们所建的来突出场地的戏剧性！是的，这是很特殊的一个。

The House Cast in Liquid Stone

It's a second home on a rocky outcrop at the start of the western ghats, Khopoli, in Maharashtra, India. As an area of high precipitation in the monsoons, and equal heat during the summers, the site changes remarkably from March to July, with the onset of the southwesterly monsoons.

The local black basalt of the region is what this site was about.

The architects chose to build the house as an accretion on this rocky basalt outcrop. It's an outgrowth which was made of a mix of water, sand, cement and the granular basalt. Concrete is finely honed to serve as refuge, to face the climatic changes that the site offered.

The house was conceived as a cast for human occupation, a refuge which trapped the views, the sun, the rain, the air, and became one with the cliff edge it stood on. Akin to the growth of a coral, the substance of the walls and roof dictate the experience of inhabiting the site.

Stone has been used in many forms, based on use, wear, grip, texture. The dark saturated mattness conjures a cool sense of refuge and calm.

Photographs cannot express the sense of weight, or the sense of release at the edge of the pool at the far end of the open terrace; the feeling of burrowing deeper en route, passes the stacked stones, to the lower bedroom.

The house, a cast, an object for living, whatever you may call it, has transformed into a belvedere to minutely observe and sense the spectacle of nature, of shade as a retreat against the sharp tropical sun, the resurgence of life, and a sudden BURST of green, when the hard pounding monsoon arrives, the waft of breezes fills the air with the fragrance of moist earth, and the movement of stars crosses the very dark night skies.

To heighten the drama of the site through what we build, without building a dramatic building! A peculiar one YES.

SPASM Design Architects

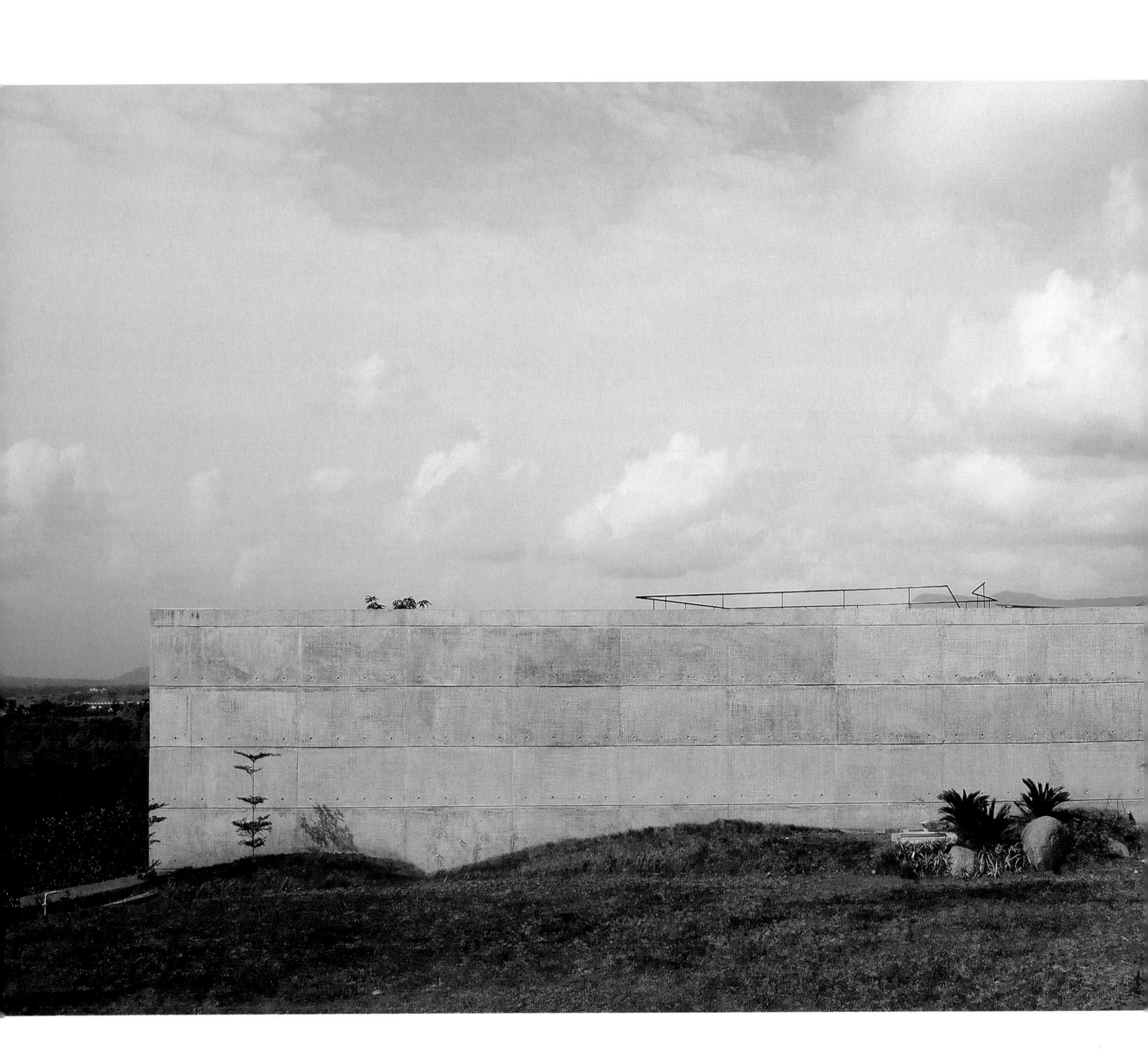

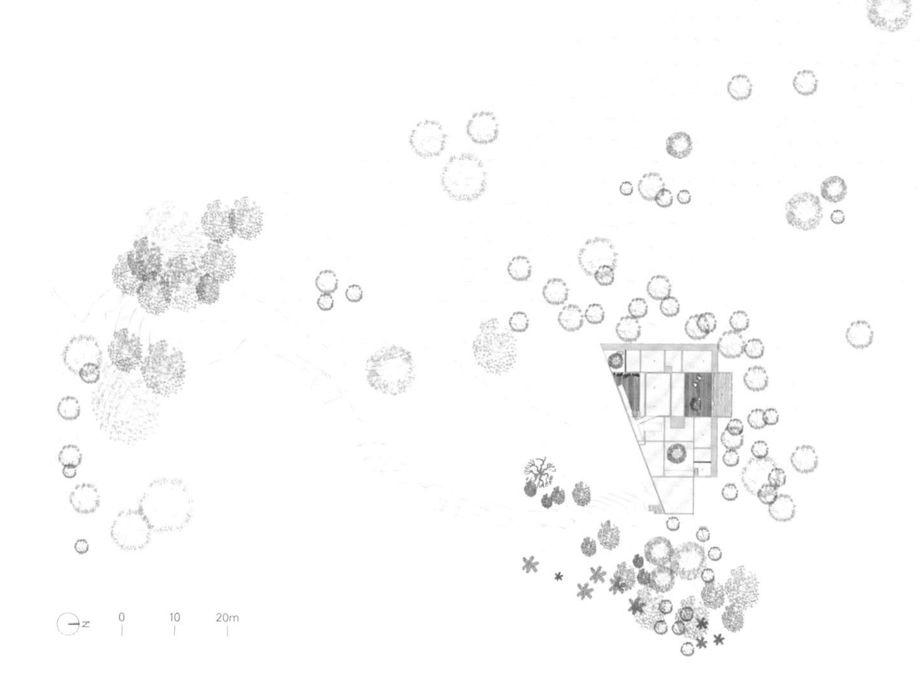
0
10
20m

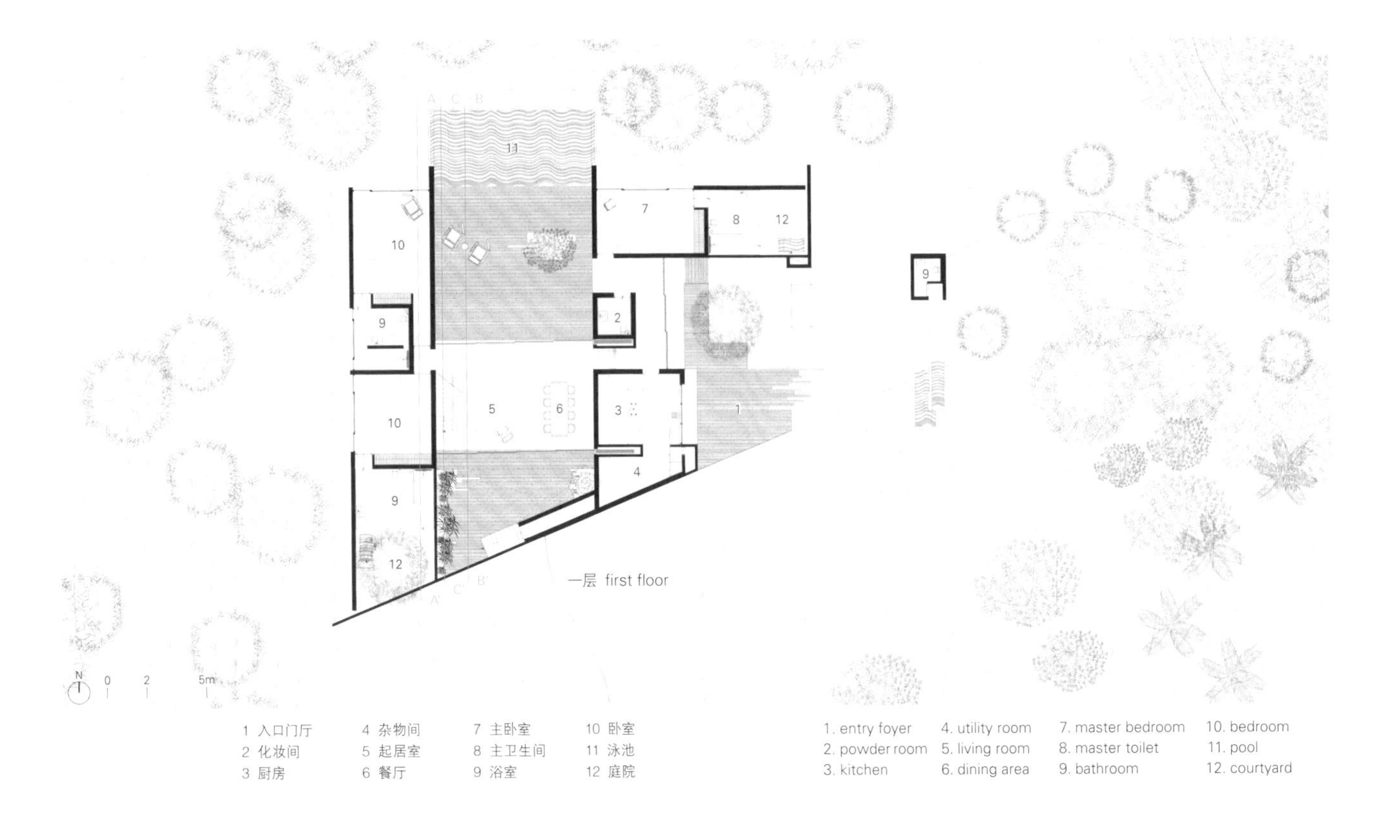

1 入口门厅
2 化妆间
3 厨房
4 杂物间
5 起居室
6 餐厅
7 主卧室
8 主卫生间
9 浴室
10 卧室
11 泳池
12 庭院

1. entry foyer
2. powder room
3. kitchen
4. utility room
5. living room
6. dining area
7. master bedroom
8. master toilet
9. bathroom
10. bedroom
11. pool
12. courtyard

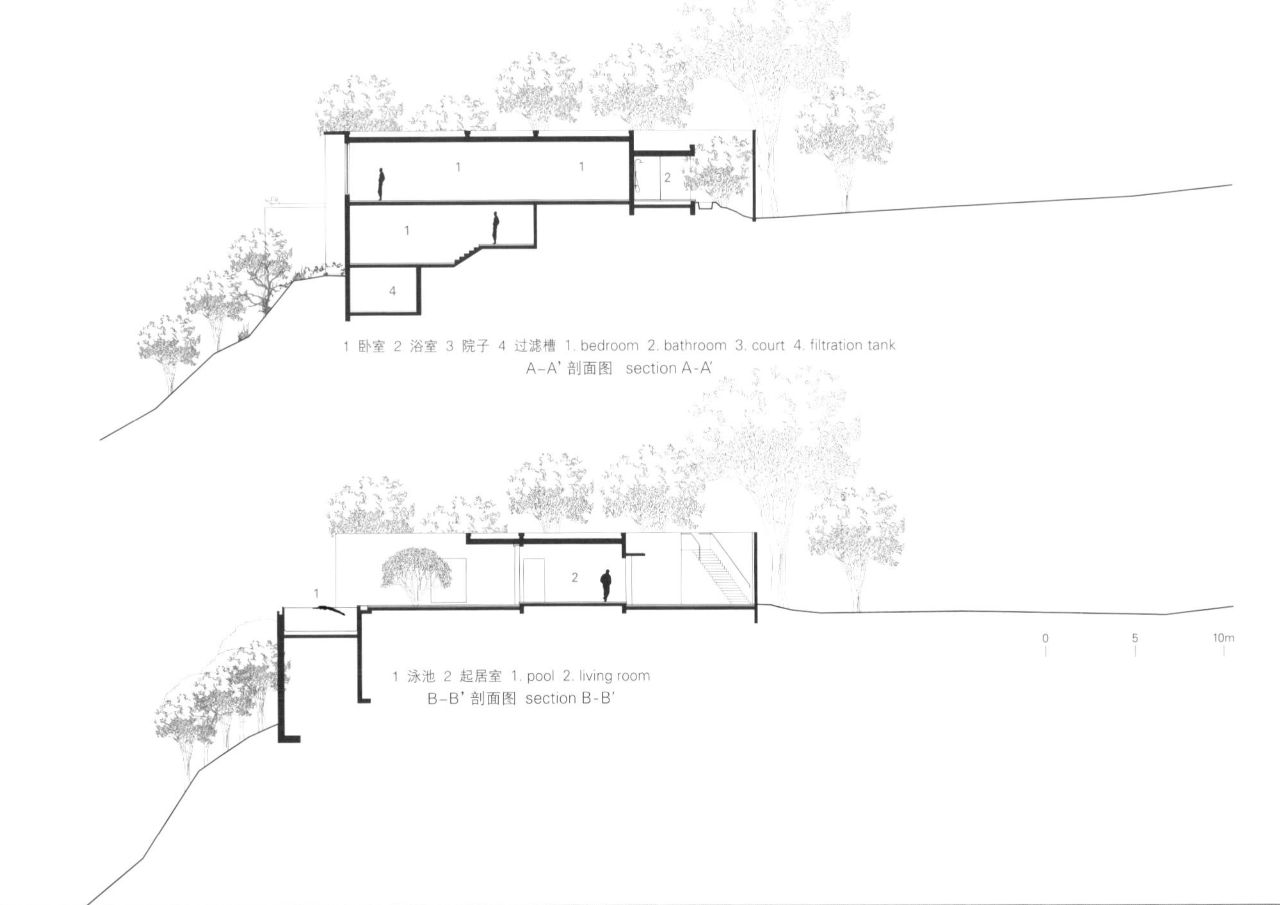

1 卧室 2 浴室 3 院子 4 过滤槽 1. bedroom 2. bathroom 3. court 4. filtration tank
A-A' 剖面图 section A-A'

1 泳池 2 起居室 1. pool 2. living room
B-B' 剖面图 section B-B'

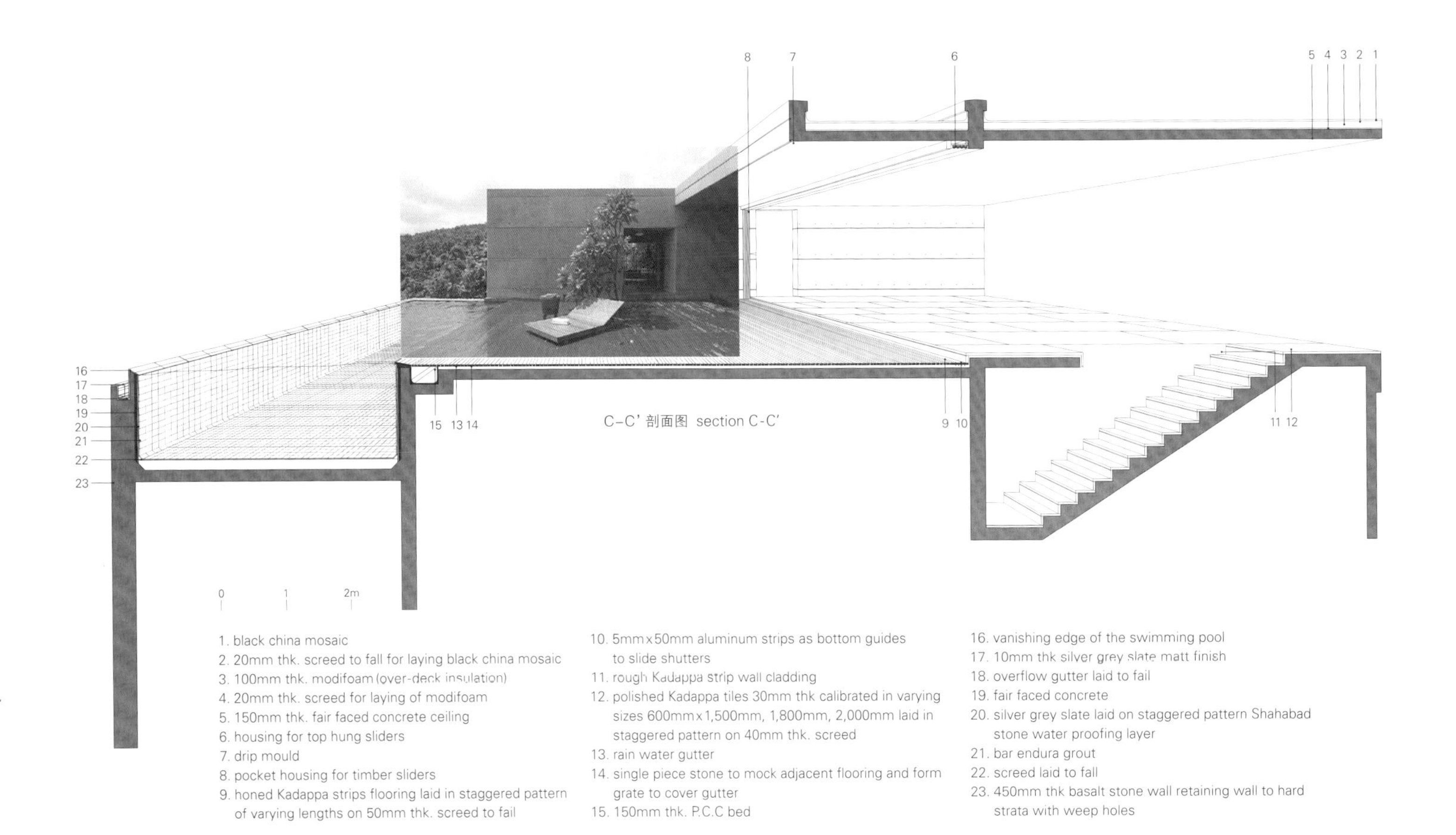

C–C' 剖面图 section C-C'

1. black china mosaic
2. 20mm thk. screed to fall for laying black china mosaic
3. 100mm thk. modifoam (over-deck insulation)
4. 20mm thk. screed for laying of modifoam
5. 150mm thk. fair faced concrete ceiling
6. housing for top hung sliders
7. drip mould
8. pocket housing for timber sliders
9. honed Kadappa strips flooring laid in staggered pattern of varying lengths on 50mm thk. screed to fail
10. 5mm x 50mm aluminum strips as bottom guides to slide shutters
11. rough Kadappa strip wall cladding
12. polished Kadappa tiles 30mm thk calibrated in varying sizes 600mm x 1,500mm, 1,800mm, 2,000mm laid in staggered pattern on 40mm thk. screed
13. rain water gutter
14. single piece stone to mock adjacent flooring and form grate to cover gutter
15. 150mm thk. P.C.C bed
16. vanishing edge of the swimming pool
17. 10mm thk silver grey slate matt finish
18. overflow gutter laid to fail
19. fair faced concrete
20. silver grey slate laid on staggered pattern Shahabad stone water proofing layer
21. bar endura grout
22. screed laid to fall
23. 450mm thk basalt stone wall retaining wall to hard strata with weep holes

courtesy of the architect

项目名称：The House Cast in Liquid Stone
地点：Khopoli, Maharashtra, India
建筑师：SPASM Design Architects
项目团队：Sangeeta Merchant, Mansoor Kudalkar, Gauri Satam, ekha Gupta, Sanjeev Panjabi
结构工程师：Rajeev Shah & Associates / 承包商：IMPEX Engineers
用地面积：19,950m² / 有效楼层面积：638m² / 设计时间：2009.11—2010.10
施工时间：2011.5—2013.5
摄影师：©Sebastian Zachariah (courtesy of the architect)(except as noted)

古梅斯苏别墅坐落在博德鲁姆的Gümüslük村，博德鲁姆是一个有着4000年历史和文化遗产的宁静港湾。创造一个宁静的、周围是令人印象深刻的自然的统一体是该建筑方案的基础。整个过程的各个部分对时下进行了解读，它混合了地方建筑元素和一种新的设计语言。其基本理念是严格地在一个正方形中定义一个独栋建筑单元，在此区域以同样的方式建造五个单元，然后其余部分留给自然。

组成一个独栋单元的房间作为独立的体块，没有相互连接，以在中间形成半开放空间。每座住宅内都设有休息室、卧室和浴室单元，它们也彼此相互独立。这些单元之间的开放空间让人想起了荫蔽且狭窄的后街，可以让凉爽的微风通过。房屋的开放式庭院、游泳池和露台阳台被构想为施工的内在部分。当所有的部分都聚集在一起时，它们形成了一个独特的四边形，且被竹蔗工作棚所遮蔽。

石墙和宽玻璃表面以最简单和时尚的方式定义着不同区域的住宅。现有树木的位置是场地布局主要考虑的一个因素。特殊的海景和景观也对布局起着决定性的作用。建筑师主要考虑的是利用传统方法来产生能效。厚石头墙壁和双墙壁提供足够的保温。为了保持房屋内部凉爽，竹凉棚成为第二个保持凉爽的因素。半开放的流线领域让微风吹到了起居空间。

几乎所有的材料都是周边传统建筑所采用的当地天然材料。

Gumus Su Villas

Gumus Su Villas are located over the hills of Gümüslük Village, Bodrum, a tranquil harbour with over 4000 years of historical and cultural heritage. The urge to create a tranquil unity with the impressive nature formed the basis of the architectural approach. The composition that came out of the process is an up-to-date interpretation that blends elements of the local architecture with a new language of design. The basic idea was to define the single house unit strictly within a square, repeat it five times within the site and then leave the rest of the site to the nature.

The rooms that make up a single unit are detached as individual blocks so as to create semi-open spaces in between. The lounge, bedrooms, and bathrooms units that comprise each house are presented as blocks that are independent of each other. The open spaces between these units are reminiscent of shady narrow back-streets that allow cooling breezes to pass through. The houses' open courtyard, swimming pool and sun terrace were

古梅斯苏别墅

Cirakoglu Architects

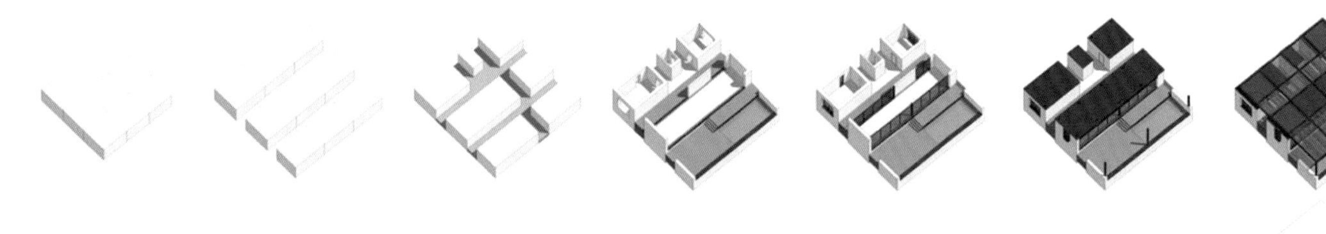

conceived as intrinsic parts of the construction. When all the parts are brought together, they form a distinct quadrangle and are shaded by a bamboo cane-work canopy.

The stone walls and the wide glass surfaces define the different areas of the house in the most simple fashion. The location of the existing trees was one of the major consideration for the site layout. The exceptional view of the sea and landscape also guided the layout decisions. Primary concern was to make use of traditional methods to provide energy efficiency. Thick stone walls and double walls provide sufficient thermal insulation. The bamboo shelter creates a secondary shading element in order to help to keep interiors cool. The semi open circulation areas let the breeze into the living spaces.

Almost all materials are local and natural materials which are used in the surrounding traditional buildings.

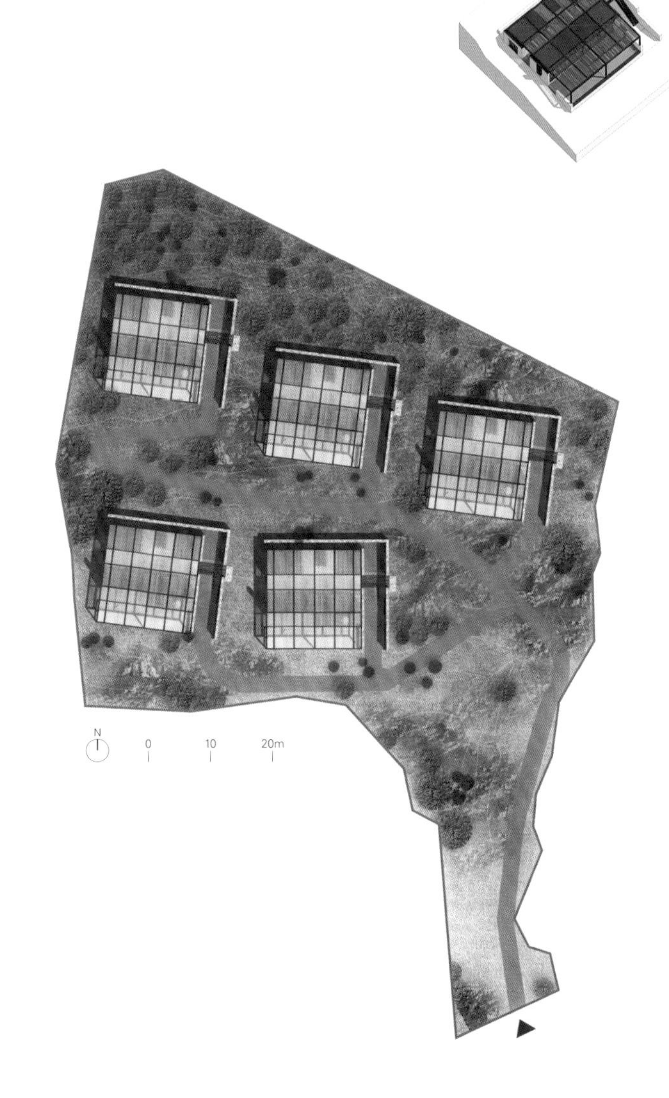

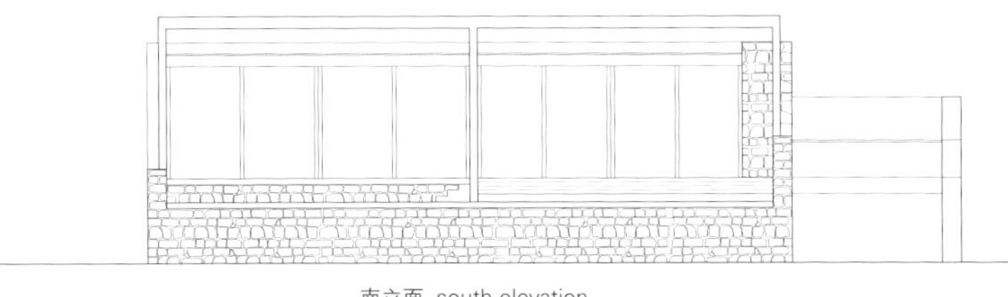

南立面 south elevation

西立面 west elevation

北立面 north elevation

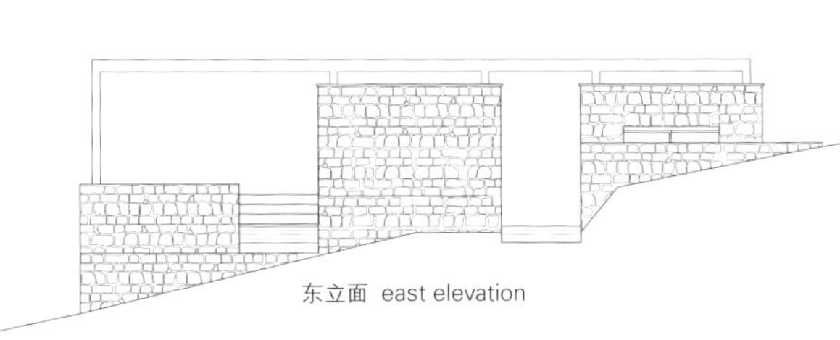

东立面 east elevation

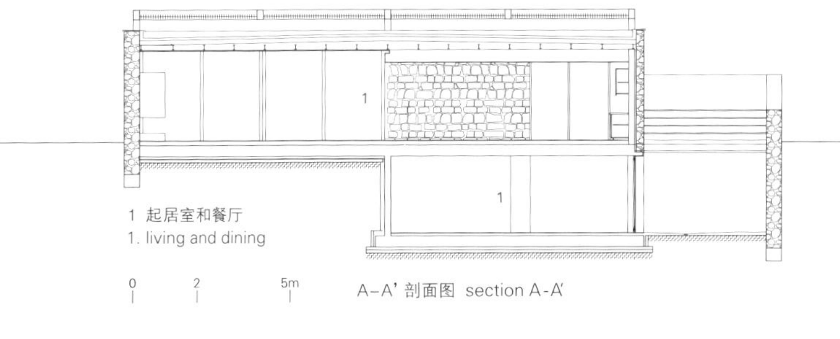

1 起居室和餐厅
1. living and dining

0 2 5m A–A' 剖面图 section A-A'

项目名称：Gumus Su Villas
地点：Gümüsluk, Bodrum, Turkey
建筑师：Cirakoglu Architects
设计团队：Alisan Cirakoglu, Ilgin Avci, Deniz Yazici, Asli Ingenc, Digdem Angin
开发商：Demirden Design
用地面积：8,000m²
总建筑面积：1,250m²
有效楼层面积：1,800m²
设计时间：2013 / 竣工时间：2014
摄影师：©Cemal Emden (courtesy of the architect)

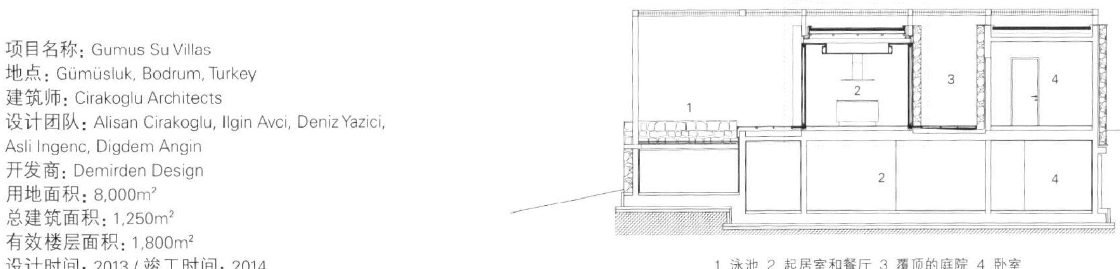

1 泳池 2 起居室和餐厅 3 覆顶的庭院 4 卧室
1. swimming pool 2. living and dining 3. sheltered courtyard 4. bedroom
B–B' 剖面图 section B-B'

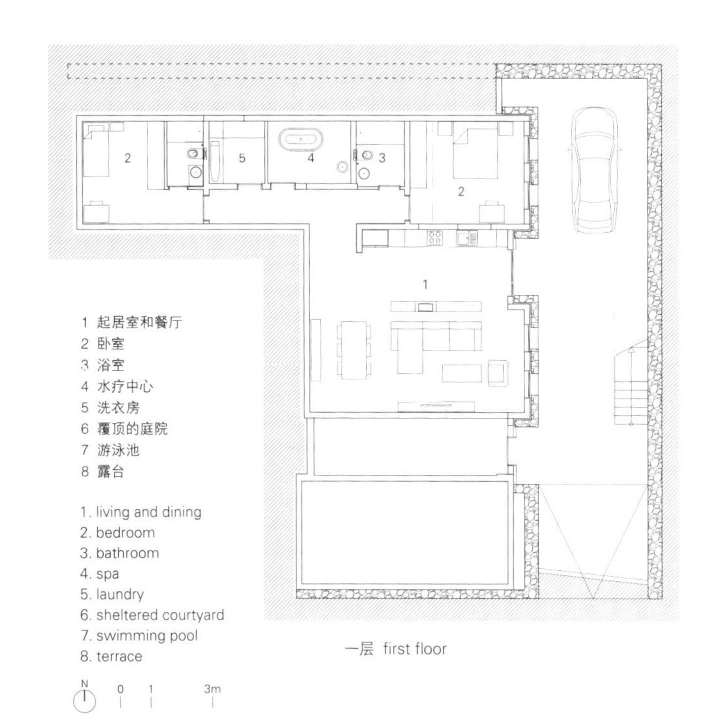

一层 first floor

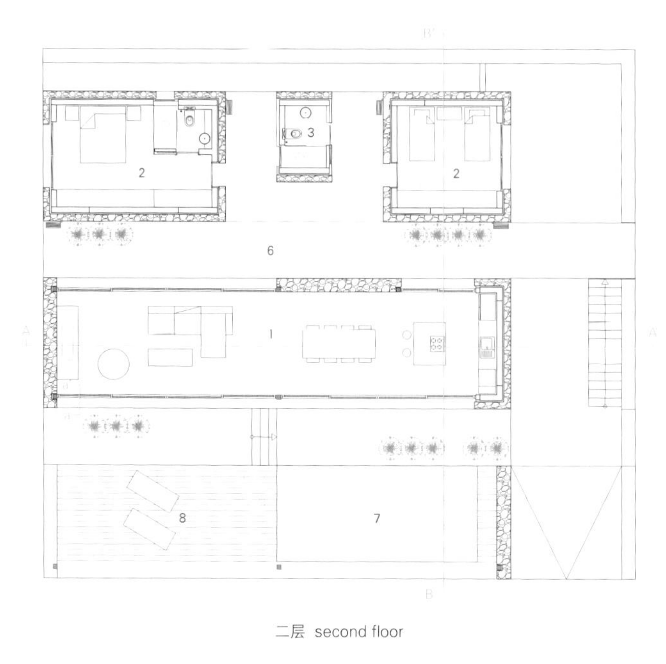

二层 second floor

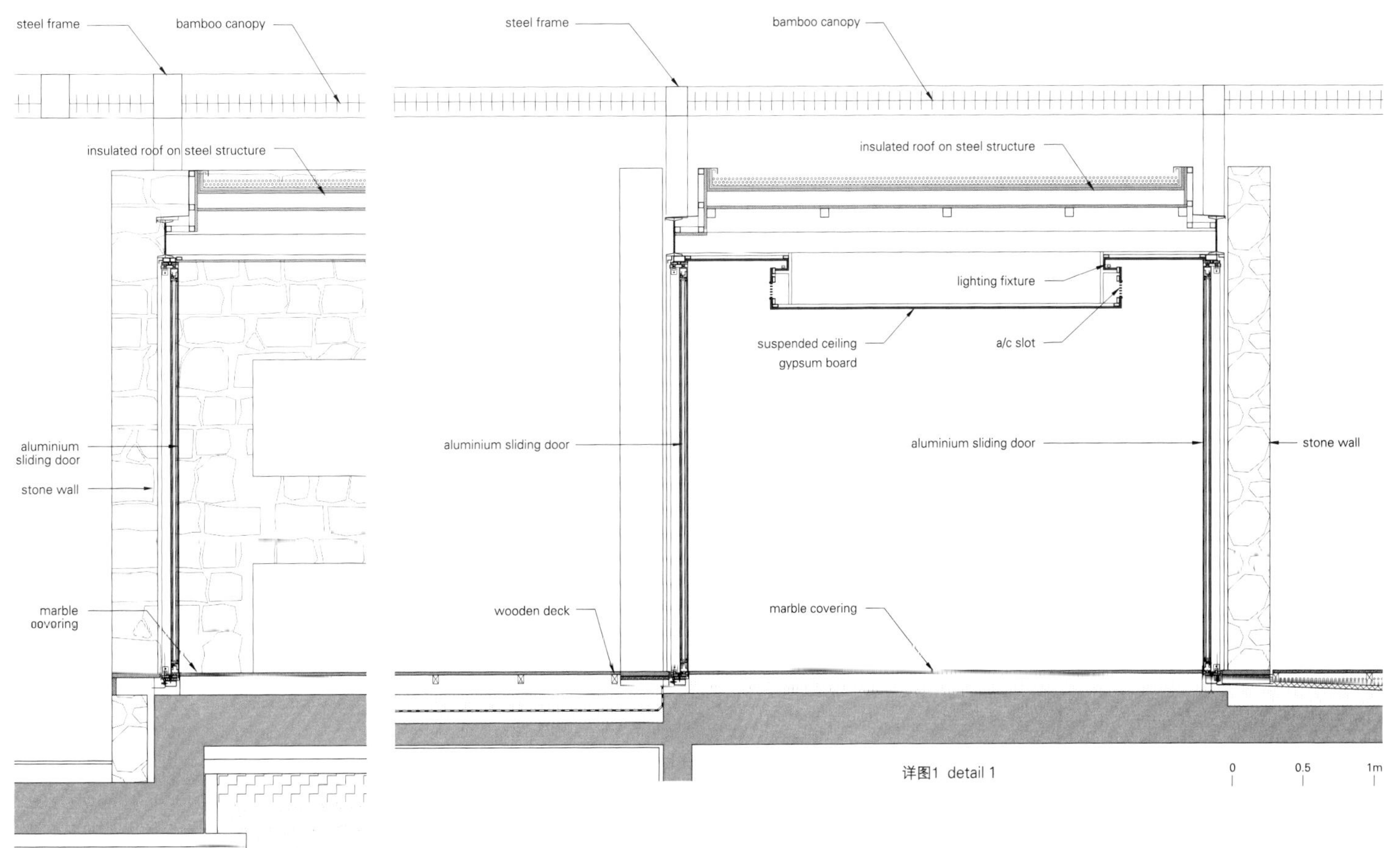

a–a' 剖面图 section a-a'

详图1 detail 1

Angelos Psilopoulos

Studied architecture at the School of Architecture, Aristotle University of Thessaloniki(AUTh), then moved on to his Post-Graduate studies at the National Technical University in Athens(NTUA). Is currently pursuing his Ph.D. at the NTUA on the subject of Theory of Architecture, studying gesture as a mechanism of meaning in architecture. Has been working as a freelance architect since 1998, undertaking a variety of projects both on his own and in collaboration with various firms and architectural practices in Greece. Since 2003, he has been teaching Interior Architecture and Design in the Department of Interior Design, Decoration, and Industrial Design at the Technological Educational Institute of Athens(TEI).

Paula Melâneo

Is an architect based in Lisbon. Graduated from the Lisbon Technical University in 1999 and received a master of science in Multimedia-Hypermedia from the cole Supérieure de Beaux-Arts de Paris in 2003. Besides the architecture practice, she also focuses on her professional activity in the editorial field, writing critics and articles specialized in architecture. Since 2001, she has been part of the editorial board of the Portuguese magazine "arqa–Architecture and Art" and the editorial coordinator for the magazine since 2010. Has been a writer for several international magazines such as FRAME and AMC. Participated in the Architecture and Design Biennale EXD'11 as an editor, part of the Experimentadesign team.

>>156

Pitsou Kedem Architects

Was established in 2002 by Pitsou Kedem and today employs nine architects. Pitsou Kedem graduated from the Architectural Association School of Architecture in London. Is a mentor of final projects at the Technion Haifa's Faculty of Architecture. Works on private as well as residential projects and recently designed commercial projects as restaurants, showrooms and hotels. Received five awards in the Israeli "Design Award" competition in the past two years. Tries to realize minimalist architecture that is based on modernism principles, architecture that related to environment and user. Is always in a search of unique materials and unconventional solutions.

©FG+SG Architectural Photography

left _ Carlos Castanheira, right _ Álvaro Siza

>>118

Álvaro Siza

Is a Portuguese architect, awarded the Pritzker Prize in 1992. Is widely regarded as one of the great living masters of architecture. Siza's works are often presented in simple but elegant forms and his constructions are highly poetic. His precise control over geometry, space and details, and his careful reflection upon the environment and scale of building works have become his defining traits. The simple minimalism of his architectural vocabulary is linked closely to the landscape, and coupled with a deep respect for region, culture and history, gives his work an irresistible tension and vitality.

Carlos Castanheira

Was born in Lisbon, 1957. Graduated in architecture from the School of Fine Arts in Porto in 1981. Lived in Amsterdam from 1981 to 1990 where he worked as an architect and attended the course of architecture in the Academie voor Bouwkunst van Amsterdam. Has been collaborating with Álvaro Siza since he was a student. In 1991, he set up a private practice with Maria Clara Bastai. Their work to date has been primarily in the private sector. Within the architectural field, his work has extended to curating and organizing exhibitions, editing and publishing boos and catalogues.

>>166

SPASM Design Architects

Was established in 1995 by two principals Sanjeev Panjabi(1969) and Sangeeta Merchant(1967). Both were educated at the Academy of Architecture, Mumbai(1987~1992). SPASM was formed when two of them got together for one project way back in 1997. Since 1987, they shared ideas, traveled together and begun to understand each other's divergent thought processes and approach to real life situations. The firm received the AR Awards 09 for the Commendation for EXIM Tower. Is busy with the construction and design of several residential, commercial and mixed use projects in India and Tanzania.

>>100

Herzog & de Meuron

Was established in Basel, 1978. Has been operated by senior partners; Christine Binswanger, Ascan Mergenthaler and Stefan Marbach, with founding partners Pierre de Meuron and Jacques Herzog. Has designed a wide range of projects from the small scale of a private home to the large scale of urban design. While many of their projects are highly recognized public facilities, such as their stadiums and museums, they have also completed several distinguished private projects including apartment buildings, offices and factories. The practice has been awarded numerous prizes including "The Pritzker Architecture Prize" in 2001 and the "RIBA Royal Gold Medal" in 2007.

>>88

MVRDV

Was set up in Rotterdam, the Netherlands in 1993 by Winy Maas, Jacob van Rijs, and Nathalie de Vries. Engages globally in providing solutions to contemporary architectural and urban issues. Realized projects including the Dutch Pavilion for the World EXPO 2000 in Hanover. The work of MVRDV is exhibited and published worldwide and receives international awards. Together with Delft University of Technology, MVRDV runs "The Why Factory", an independent think tank and research institute providing argument for architecture and urbanism by envisioning the city of the future.

>>70

Emre Arolat Architects

The founding partner of Emre Arolat Architects, Emre Arolat was born in 1963. Throughout his career, he has been participating in various conferences and seminars. He has also curated some exhibitions including the first Istanbul Design Bienniale. Has taught at several architectural schools in Turkey and abroad, including Mimar Sinan Fine Arts University, Istanbul Bilgi University, Berlage Institute for Architecture, Delft University of Technology. Became a visiting juror at Pratt Institute and Middle East Technical University. Has received many national and international awards including Aga Khan Award for Architecture, Mies van der Rohe Award and AR Awards for Emerging Architecture.

>>28

Adam Khan Architects

Adam Khan was born in London and graduated from the London Metropolitan University, studying with Florian Beigel. He established Adam Khan Architects in 2006. The practice has received numerous awards including Public Building Architect of the Year in 2012 and 3 RIBA National awards. New commissions include Ellebo Housing Estate in Copenhagen, and new social housing in Hamburg and London.

>>178

Cirakoglu Architects

Was founded by Alisan Cirakoglu. After graduation in 1996 and having a master's degree from Middle East Technical University, he worked with several practices in Ankara and Istanbul. In 2002, he founded Cirakoglu Architects in Istanbul. He has been awarded the Arkitera Young Architect Award in 2008. He was a jury member for the National Architecture Awards 2012, Mimed 2010, Archiprix Turkey 2010 as well as several architectural competitions.

>>40

Agence Guinée Et Potin Architectes

Was established by Anne-Flore Guinée and Hervé Potin in Nantes, 2002. Anne-Flore Guinée was born in 1973 in Saint Nazaire. Graduated from the Architecture School of Britany in 1997. Currently is teaching at the Horticultural Institute of Angers. Hervé Potin was born in 1972 in Orléans. Graduated from the Architecture School of Britany in 1996. Is a professor of the Architecture School of Nantes.

Both in objects, designs and architecture drawings, they select a digest of representations of ingenuous or infantile taste. Certainly, the infantile effect appears also in their architecture.

>>10

Amateur Architecture Studio

Wang Shu is a Chinese architect based in Hangzhou. Received a B.S. in 1985 and a master's degree in 1988 from Nanjing Institute of Technology, Department of Architecture. Founded Amateur Architecture Studio in 1997 with his wife Lu Wenyu and received a Ph.D from Tongji University, School of Architecture in 2000. Since 2007, he has been a dean of Architecture School in China Academy of Art in Hangzhou. In 2012, he became the first Chinese citizen to win the Pritzker Prize.

>>144

standardarchitecture

Was founded by Zhang Ke in 2001. Is a leading new generation design firm engaged in practices of planning, architecture, landscape, and product design. Based on a wide range of realized buildings and landscapes in the past ten years, it has emerged as the most critical and realistic practice among the youngest generation of Chinese architects and designers.

Zhang Ke received B.Arch and M.Arch from School of Architecture at the Tsinghua University. In 1998, he received another M.Arch at the Graduate School of Design, Harvard University. Currently he lectures at various Universities including Tsinghua University and University of Hong Kong(HKU). Was awarded Design Vanguard of the World from the Architecture Record in 2010.

>>52

Oto Arquitectos

Was founded in 2008 by four university colleagues; André Castro Santos, Miguel Ribeiro de Carvalho, Nuno Teixeira Martins and Ricardo Barbosa Vicente in Lisboa, Portugal. The practice engages globally in providing solutions for the future in both architectural and urban issues. Their approach is highly collaborative, multidisciplinary and universal, research-based design method involves clients, and experts from a wide range of fields from early on in the creative process. They embrace all projects with the eager to discover something new, exiting for their clients. They are now working in Angola, Cabo Verde, Mozambique, Portugal, São Tomé and searching for new opportunities worldwide.

图书在版编目(CIP)数据

重塑建筑的地域性 : 汉英对照 / 韩国C3出版公社编 ; 刘懋琼, 陈玲译. —大连: 大连理工大学出版社, 2014.12
(C3建筑立场系列丛书)
书名原文: C3 Re-assessing Local Identity
ISBN 978-7-5611-9638-0

Ⅰ. ①重… Ⅱ. ①韩… ②刘… ③陈… Ⅲ. ①建筑设计—汉、英 Ⅳ. ①TU2

中国版本图书馆CIP数据核字(2014)第278699号

出版发行：大连理工大学出版社
（地址：大连市软件园路 80 号　　邮编：116023）
印　　刷：上海锦良印刷厂
幅面尺寸：225mm×300mm
印　　张：12
出版时间：2014 年 12 月第 1 版
印刷时间：2014 年 12 月第 1 次印刷
出 版 人：金英伟
统　　筹：房　磊
责任编辑：张昕焱
封面设计：王志峰
责任校对：周　一

书　　号：978-7-5611-9638-0
定　　价：228.00 元

发　行：0411-84708842
传　真：0411-84701466
E-mail：12282980@qq.com
URL：http://www.dutp.cn